ECOPHYSIOLOGY AND POSTHARVEST QUALITY OF SALAK

(*Salacca zalacca* (Gaertn.) Voss)

Reni Lestari

Gedruckt mit Unterstützung des Deutschen Akademischen Austauschdienstes

Bibliografische Information Der Deutschen Bibliothek

Die Deutsche Bibliothek verzeichnet diese Publikation in der Deutschen Nationalbibliografie; detaillierte bibliografische Daten sind im Internet über http://dnb.ddb.de abrufbar.

ISBN 3-8325-0850-3

Logos Verlag Berlin
Comeniushof, Gubener Str. 47,
10243 Berlin
Tel.: +49 030 42 85 10 90
Fax: +49 030 42 85 10 92
INTERNET: http://www.logos-verlag.de

ACKNOWLEDGEMENTS

The author is deeply indebted to the supervisors, Dr. Georg Ebert and Dr. Susanne Huyskens-Keil, for their guidance, scientific and moral support during the study that will positively influence the author's academic life.

Sincerely grateful to Dr. Detlef Ulrich (Institut for Plant Analysis, Federal Centre for Breeding Research on Cultivated Plants (BAZ), Quedlinburg, Germany) for permitting and assisting me in working on aroma compound at his facilities. Special thank for kind collaboration in aroma compound study is given to Prof. Dr. C. Hanny Wijaya (Department of Food Technology and Human Nutrition, Bogor Agricultural University, Bogor, Indonesia).

Appreciations are dedicated to all colleagues in the Institute for Horticultural Science, Humboldt University Berlin. Most particularly to Dr. R. Feyerabend, Ms. K. Mühe and S. Block for technical help during my study; Ms. R. Lindemann, Ms. G. Nowottnick, Ms. I. Dressel, Ms. I. Gründel, Ms. M. Heller, Ms. U. Twardąwski and T. Schober for their help in laboratory analysis; as well as Ms. P. Smith, Dipl. Ing. K. Zippel, Dipl. Ing. S. Schäfer, Dr. V. Fernandez, Dr. S. Müller and Dipl. Ing. I. Eichholz for their support and generous.

Thanks are also given to colleagues and family in Indonesia who assisted me during the field works in Yogyakarta province of Indonesia, especially Dr. Taryono, Dra. Ratih S.S., MS., Ir. Probo and Drs. Syaifurrahman, SH.

Deeply appreciation is expressed to DAAD (German Academic Exchange Services) in Bonn, Center for Plant Conservation – Bogor Botanical Gardens, Indonesian Institute of Science and the Indonesian Government for the financial supports and good coordination.

Finally, this publication is devoted to my husband, Noer Azam Achsani, to our children Faaiz and Farhan, our parents, our sisters and brothers for their help, patience and tolerance.

Berlin, February 2005

Reni Lestari

Table of contents

List of Tables

Page

List of Figures

Page

Abbreviation

AAS	:	Atomic Absorption Spectophotometry
ADF	:	Acid Detergen Fibre
AIS	:	Alcohol Insoluble Solid
ANOVA	:	Analysis of Variance
AOAC	:	Association of Official Analitycal Chemist
BA	:	Benzyl Adenin
BAP	:	Benzyl Amino Purin
DIN	:	Deutsche Industrie Norm
CIS	:	Cooled Injection System
DM	:	Dry Matter
DW	:	Dry Weight
EDTA	:	Ethylene Diamin Tetra Acids
EDTA-SP	:	Ethylene Diamin Tetra Acids-Soluble Pectin
FID	:	Flame Ionisation Detector
FW	:	Fresh Weight
GC-MS	:	Gas Chromatography – Mass Spectometry
GCO	:	Gas Chromatography Olfactometry
HP	:	Hewlett Packed
HPLC	:	High Performance Liquid Chromatography
IAA	:	Indol Acetic Acid
IBA	:	Indol Butyric Acid
IS	:	Internal Standart
ISO	:	International Organisation for Standardisation
ISP	:	Insoluble Pectin
LAI	:	Leaf Area Increment
LDPE	:	Low Density Polyethylene
LSD	:	Least Significant Diference
MS	:	Murashige and Skoog
NBS	:	National Bureau of Standard
NDF	:	Neutral Detergen Fibre
NIF	:	Nasal Impact Frequency
NIST	:	National Institute of Standards and Technology
PAR	:	Photosynthesis Active Radiation

PDMS	:	Poly(dimethylsiloxane)
PE	:	Polyethylene
PG	:	Polygalacturonase
PME	:	Pectinmethylesterase
PP	:	Polypropylene
PVC	:	Polyvinilchloride
SBSE	:	Stir Bar Sorptive Extraction
SE	:	Standard Error
SLI	:	Shoot Length Increment
SSC	:	Soluble Solid Content
TA	:	Titratable Acidity
TDR	:	Time Domain Reflectometry
TDU	:	Thermal Desorption Unit
v/v	:	volume/ volume
w/v	:	weight / volume
WSP	:	Water Soluble Pectin
ZMP	:	“Zentrale Mark- und Preisberichtstele”

1. GENERAL INTRODUCTION

In Indonesia, there is a large diversity of tropical fruit tree species, including banana, mango, papaya, pineapple, carambola, durian, guava, mangosteen, rambutan, sapota and salak. The Indonesian government has selected 8 outstanding fruit species according to their marketability, economical value, geographical distribution and climate suitability (Winarno, 2000). Salak is one of these candidates due to its superior characteristics. Salak species are regarded to be a very important fruit crop for the Indonesian market. In Indonesia, salak production has increased from 97.5 t in 1989 to 292.2 t in 1994 (Biro Pusat Statistik, 1995). Moreover, salak has a potentially high value for export purposes (Dendi, 1997; Djaafar, 1998). There are 30 cultivars of salak, which distribute across the Indonesian islands (Sudaryono et al., 1993; Kusumo, 1995). The availability of superior quality fruit of selected cultivars will promote the marketing of salak, such as cultivars "pondoh" from Yogyakarta and "gula pasir" from Bali provinces. The excellent taste of these cultivars is very well-known and appreciated in these regions. In regard to "pondoh", fruit production has increased from 8.6 t in 1995 to 24.3 t in 2001 (Departemen Pertanian, 2001). Fresh fruits of this cultivar have also been exported to Singapore and United Kingdom and in the near future it will be sent to Malaysia, Thailand, Hongkong and Saudi Arabia (Djaafar, 1998).

The development of a prospective fruit production is very important in order to increase the competence in responding to the demand for local and export markets. To promote intensive salak production and to assure high fruit quality, an interdisciplinary approach is required. Therefore, clarification of quality aspects of the fruit is also needed to promote the market expands.

However, the major problems of salak product in Indonesia are (1) fruit quality, (2) quantity of yield and (3) continuity of supply (Winarno, 1997). These arise from non-intensive plantation systems and seasonal production patterns (Arintadisastra, 1997). There has been an increasing interest in improving salak production techniques and postharvest properties in Indonesia. These approaches include better knowledge on pre- and postharvest properties of salak fruit. With respect to pre-harvest aspects, there is a special need for focusing on the ecophysiology of salak plant since it plays a major role in fruit quality traits. A better understanding of postharvest properties of salak will help to maintain fruit quality during storage, shipping and processing.

The major objectives of this study were:

(1) to investigate the ecophysiological requirements of different salak cultivars,

(2) to investigate postharvest quality aspects of salak fruits,

(3) to assist the establishment of a quality-oriented production and postharvest management of salak fruits.

In this thesis, the results of the studies will be presented as follows:

In Section 2, a comprehensive review on salak plant and fruit is conducted. The information regarding botanical aspects as well as current information on salak production techniques and postharvest properties, with special emphasis on "pondoh" cultivars, is presented.

In Section 3, responses of 4 salak cultivars, "pondoh super", "pondoh hitam", "pondoh manggala" and "gading" to different growing media will be given. Three different growing media, peat, sand and sand/compost mixture heve been used for the study. The responses observed include growth, net CO_2 assimilation rate, leaf colour and leaf mineral content of salak seedlings.

In Section 4, the effects of light intensity and water supply on growth, net CO_2 assimilation rate and leaf mineral content of salak seedlings are presented.

In Section 5, fruit quality parameters of salak cultivars "pondoh super", "pondoh hitam", "pondoh manggala" and "gading" will be compared. Quality attributes include external properties, nutritional values, sensory attributes evaluated by a consumer panel and aroma compounds.

In Section 6, physical and physiological changes during maturation and ripening of salak fruits are determined. The information will be useful for optimising the consumer-oriented fruit quality and for extending the marketing period, especially for export purposes.

The findings of this research will help to promote salak production in Indonesia, both for domestic and export markets.

2. LITERATURE REVIEW

2.1. Origin and Botany

The genus *Salacca* belongs to the subtribe *Calaminae*, the tribe *Calameae* and the subfamily *Calamoideae* of the family *Arecaceae* (Dransfield and Uhl, 1986). *Salacca* consists of 21 species and 4 varieties (Mogea, 1992), which are known as salak or "snake fruit" (Schuiling and Mogea, 1992; Kusumo, 1995; Supriyadi et al., 2002). Some local people in Indonesia called the fruit as "saloobi" (Batak, North Sumatra), "sekomai" (Jambi, Sumatera), "hakam", "toosoom" (Dayak, Kalimantan) and "sala" (Sulawesi) (Tjahjadi, 1989). The genus *Salacca* is indigenous throughout the Indo-Malaysian region, i.e. in Indonesia, Burma (Myanmar), Thailand, Malaysia, Kampuchea, Laos, Vietnam and the Philippines (Miller, 1978). The plants are cultivated in Thailand, Malaysia, Indonesia, New Guinea, the Philippines, Queensland (Australia), Ponape Island (Caroline Archipelago) and the Fiji Islands (Miller, 1978; Schuiling and Mogea, 1992). One species of salak palm, which grows in Northern Sumatra, is ascribed to a distinct species, *Salacca sumatrana* Becc. Another species, *S. zalacca*, is cultivated elsewhere in Indonesia. The synonym of "*S. edulis*" Reinw. is sometimes used in the literature, however Mogea (1982) has corrected the name to *S. zalacca* (Gaertn.) Voss. This species is subdivided into two botanic varieties, var. *zalacca* from Java and var. *amboinensis* (Becc.) Mogea from Bali and Ambon (Schuiling and Mogea, 1992). *S. wallichiana* is cultivated only in Thailand (Polprasid, 1992; Dangcham and Siriphanich, 2001).

Salak plants can be productive for 50 years or more (Schuiling and Mogea, 1992). There are 30 *S. zalacca* cultivars grown in various production areas in Indonesia, e.g. Karangasem (Bali), Sleman (Yogyakarta), Condet (Jakarta), Batujajar (West Java), Bangkalan (Madura) and Enrekang (South Sulawesi) (Kusumo, 1995). In these areas, various cultivars are known. In Bali for example, 10 cultivars are known (Suter, 1988a), such as "gondok", "nenas", "nangka", "gula pasir", "kelapa" and "putih" (Sudaryono et al, 1993; Suter, 1988a). Other cultivars are named as "padangsidempuan", "condet", "pondoh", "suwaru", "nglumut", "madu" and "gading" (Sudaryono et al., 1993; Sudaryono et al., 1994; Kusumo, 1995). The cultivar name refers to fruit taste, peel colour, aroma, shape or the place of production. Based on the thickness of the fruit pulp, "gondok" from Bali is the preferred cultivar (Purnomo and Sudaryono, 1994), whereas in terms of high edible portion, "putih" from Bali is recommended (Suter, 1988a). On the other hand, the tastiest cultivars are "pondoh" from Yogyakarta and "gula pasir" from Bali (Purnomo and Sudaryono, 1994). Currently, salak "pondoh" has been divided into three sub-cultivars, i.e. "pondoh super", "pondoh hitam" and "pondoh manggala" (Djaafar and Thamrin, 1996; Santosa et al., 1996a; Santosa et al., 1996b). These cultivars are differentiated by the colour of leaf, stem, fruit kernels, peel and pulp, fruit size, weight and fruit taste (Santosa et al., 1996b; Djaafar, 1998).

Salak belongs to a group of palms which do not form trunks, but rather sprout their leaves from the ground level. The plant is a low, erect, heavy headed, extremely spiny palm with a maximum height of 5 m (Page, 1984) (figure 1a). Roots are born from the trunk where it comes in contact with the soil. The feather-leaves are up to 7 m long and the margins of midribs, petioles and leaflets are set with long sharp spines. Leaflets are up to 70 cm long and dark green. A detailed botanical description of salak can be found in Ochse (1931), Furtado (1949), Corner (1966), Whitmore (1973), Miller (1978) and Schuiling and Mogea (1992).

Epigeal seed germination occurs in less than one week and becomes evident when the hypocotyl dislodges the germ-pore cap and appears as an emergent plug (figure 1b). In approximately 3 months, the seedlings attain a fully expanded bifid juvenile leaf stage and remain attached to the seed, with no evidence of endosperm disintegration (Miller, 1978) (figure 1c). Seed germination includes into Richard`s admotive germination type and the Archontophoenix type of Gatin. The cotyledon does not differentiate a long connective or a cotyledonary petiole. The juvenile stage of seedling leaf fits into Tomlinson´s reduplicate palms, class four category where the first eophyll is bifid and the adult foliage leaves are paripinnate (Miller, 1978).

Figure 1a. Salak palms, b. 2-weeks-old seedling, c. 8-months-old seedling, d. Female inflorescence, e. Male infloresence, f. Fruit with seed (upper) and pulp segments (lower)

The salak palm is *dioecious* (Ochse et al., 1961; Purseglove, 1968, Mogea, 1978, Westphal and Jansen, 1989). However, some palms planted in Bali are reported as being *monoecious* (Mogea, 1978; Moncur and Watson, 1987; Santoso, 1990; Schuiling and Mogea, 1992; Rukmana, 1999). Pollen of three cultivars salak from Bali, i.e. "gula pasir", "gading" and "boni" was known to be sterile. Seed of those Bali cultivars was apomictic and includes in the category of "adventitious embryony" (Hutauruk, 1999). The inflorescence bears both hermaphrodite and staminate flowers; the latter produce functional pollen. Female flowers are larger than the males, and derive from shorter spadices of 20 to 30 cm length, borne 1 to 3 together. The flowers are in pairs in the axils of the scales; the three petals have a shiny dark red colour (Ochse, 1931) (figure 1d). Male flowers are closely packed in finger-like spadices of 50 to 100 cm length, borne 1 to 12 together in bunches (figure 1e). The flowers develop in pairs in the axils of the scales, they are reddish and calyx and corolla are three parted with six stamens. The development of flower buds in the genus *Salacca* has been described by Fisher and Mogea (1980). Flower buds are enclosed in a chamber within the leaf base and expand through a slit on the abaxial surface of the leaf base. The inflorescence bud is an axillary meristem that becomes radially displaced by adaxial growth of the leaf primordium (Fisher and Mogea, 1980). Inflorescences are formed throughout the year, however the main flowering period is from August to October (Moncur and Watson, 1987).

Van Heel (1977) reported on the development and morphology of the ovules in the genus *Salacca*. Ovules possess a small nucleus which is soon absorbed by the young embryo sack. There are two free integuments, both undergo extensive growth in the course of seed formation, whereas the inner integument is entirely absorbed by the young embryo sack. The outer integument continues its growth in thickness and ultimately forms a sarcotestae in the ripe seed. The edible part of the *Salacca* fruit is an aril in a loose functional sense only, not in a precise morphological context (Van Heel, 1977). In the ripe fruit, the sarcostestae reveal the same vascular bundle pattern as is present in the outer integuments of the young seeds. The hard kernels are formed by the endosperm.

Fruits are located in tight, globose bunches, round, 2.5 - 10 cm x 5 - 8 cm across. They are covered with regularly arranged scales developing from the peel of the fruit (pericarp) giving it the appearance of a snake or reptiles'skin. The aromatic fruits enclose a soft, transculent pulp whith a taste comparable to a combination of apple, pineaple and banana (Schuiling and Mogea, 1992). Fruits contain 1 to 3 blackish kernels, which are about 1 cm in diameter (Ochse, 1931) (figure 1f).

The fruit type is not easy to categorise. Martius (1823, 1850) and Blume (1836) cited by Miller (1978) refer the fruit being a berry, Ochse (1931) and Backer and Brink (1968) reported it as drupe, Bailey (1946) called it "drupe-like", Corner (1966) cited by Miller (1978) described it as "drupe without stone" and Miller (1978) assumed it as "a few-seeded berry" or "not a true berry".

2.2. Cultivation Practices

Salak is native to the lowlands of humid tropical regions (Schuiling and Mogea, 1992) (figures 2 and 3). The variation of important environmental conditions for salak cultivation is such as 0 – 700 m altitude with an average rainfall of 2400 – 4800 mm per year (Tjahjadi, 1989; Santoso, 1990). Salak plants are distributed at 50 – 300 m altitude in areas with a rainy season of 5 – 7 months per year or at 450 – 650 m altitude with rainy season of 8 - 12 months (Sudaryono, 1995). Because of its superficial root system, salak requires a high ground water table, well distributed rainfall or irrigation during the dry period of the year (Tjahjadi, 1989; Santoso, 1990; Schuiling and Mogea, 1992), but it does not tolerate flooding (Schuiling and Mogea, 1992). The amount of water supply to salak is critical, since heavy rainfall causes flower bud decay (Mogea, 1979; Santoso, 1990). On the other hand, dry periods may result flower bud drying-off (Mogea, 1979) and less shoot growth for air-layering propagation purposes (Santoso, 1990). Fruit yield is usually high in areas with 150 – 800 mm rainfall per month when the increase in precipitation from one month to the next is not more than 400 mm (Mogea, 1979). In dry seasons, salak palms should be supplied with 20 l water per plant at 20 day intervals (Soleh et al., 1995).

Appropriate temperature for salak cultivation is between 20 °C and 32 °C (Tjahjadi, 1989; Santoso, 1990; Soedaryono, 1994; Santosa et al, 1996a). Light intensity should be from 30 % to 80 % of full sun light (Tjahjadi, 1989; Santoso, 1990; Sudaryono et al.,

1992). Therefore, salak palms need shading plants, such as coconut (*Cocos nucifera* L.), lanseh (*Lansium domesticum* Correa), *Leucaena glauca* Auct., jackfruit (*Artocarpus integra* Merr.), *Canangium odoratum* Baill (Tjahjadi, 1993), banana (Sudaryono et al., 1997) or mango (Schuiling and Mogea, 1992). Salak has a shallow root system and it requires wind shelter or tree rows (Santoso, 1990).

Tjahjadi (1993) reported that the soil type for salak could be sand or sandy clay or clay. Although salak is unpretentious, nutrient supply is needed. Salak requires a free-draining soil with a high organic content (Santoso, 1990; Tjahjadi, 1993; Kusumo et al., 1995). Soil pH may vary over a wide range from 4.5 to 7.5 (Santoso, 1990; Tjahjadi, 1993; Kusumo et al., 1995). Salak seedlings can grow in low pH and nutrient deficient peat media, mixed with hull of rice, if they will be supplied with foliar fertilisers (Kusmiba, 1994).

Appropriate soil and water management, pruning system and fertiliser application are prerequisites for producing high fruit quality in salak (Thamrin et al., 1998). The application of leaf sheet pruning, fertilisation, irrigation, the presence of pollinators (10 *Curculionidae* per flower bunch) and fruit set control increase the frequency of main harvests from 2 to 3 times per year (Sudaryono et al., 1999). Irrigation is necessary during the dry season. A salak palm should be watered with 20 l in intervals of 20 days (Soleh et al., 1995). The addition of 5 kg organic soil conditioner per plant was recommended to improve the irrigation efficiency. On the other hand, the use of 2 kg biofert soil conditioner per plant retarded water infiltration into the soil (Thamrin et al., 1998). Availability of macronutrients such as K, Mg and S has been reported to be a limiting factor for salak "pondoh" and other cultivars in Bali (Kusumainderawati and Soleh, 1995). B and Zn deficiency in salak have been described by Soleh et al. (1993). Pruning of salak leaves is very important especially during rainy periods to protect flower buds from decay and leaf diseases caused by *Pestalotia* sp. (Soleh et al., 1993) (figure 4). Sixteen leaves per plant for salak from Bali and 14 leaves per plant for "pondoh" are recommended (Soleh et al., 1993).

Figure 2. Typical salak plantation, with paddy field (in the front) and coconut plants (in the back)

Figure 3. Salak plantation beside a village street

Figure 4. Pruning of leaves

Figure 5. Propagation by rooting of shoots ("air layering")

Figure 6. Thinning of salak fruits

Figure 7. Harvest of salak fruits

2.3. Propagation

2.3.1. Generative propagation

The seeds for propagation purposes have to be use immediately after removal from the fruit. Seeds will germinate readily within less than one week under humid and shady conditions (Schuiling and Mogea, 1992). Sixty to ninety days after sowing, the first complete bifid leaf, about 20 to 30 cm long, will be expanded, the seedling still being firmly attached to the seed. Seeds can be sown directly in the field (2-5 seeds together in 5 cm deep holes) or in nursery beds. Seedling palm trees should be transferred to the field during the rainy season when they are about 6-months-old (Tjahjadi, 1993). The final distance between the seedlings in the field is usually 2 m x 2 m (Santoso, 1990; Tjahjadi, 1993). Purbiati et al. (1999) reported that only 10 % of "gondok" seedlings were damaged after being transferred from the nursery to the field. This generative propagation method covers the problem of an unknown sex type of the seedlings, which start first flowering after 3 to 4 years, and of a non uniform fruit quality. Hadi (2001) reported that in respect to the number of chromosomes, there are no differences between male, female and hermaphrodite salak plants, i.e. 2n = 28. In contrast, the structure and the length of the chromosomes are different between the sexes. Seeds, which show a "belt" like shape during germination, tend to develop female plants with a 77 % probability, whereas seeds without a "belt" shape develop to male plants with a 85 % probability (Hadi, 2001).

The moisture content of the seed plays an important role for its viability. After removal from the fruit, the seeds quickly loose their viability, probably because of the high irreversible water loss of the embryo. Germination rate of only 55 % was found after 1 week of storage and after 2 weeks seeds completely failed to germinated (Schuiling and Mogea, 1992). The decrease in seed moisture content from 56.7 % to 41 % reduces the germination rate from 90 % to 0 %, whereas 50 % germination correlated to a moisture content of 47.5 % (Purwanto et al., 1988). Salak seeds have no tolerance to desiccation and they are not suitable for *ex situ* cryopreservation purposes (Dickie et al., 1993).

2.3.2. Vegetative propagation

Vegetative propagation can be done by removing rooted lateral offshoots from older plants. Best results are obtained when air-layering is conducted at 3 to 6 months old shoots. The base of the offshoots, which is still attached to the parent plant, is placed in bamboo pots (10 cm in diameter), filled with a 1:1 mixture of soil and manure (Purbiati et al., 1994). Currently, the traditional bamboo pots are replaced by used infusion plastic bottles from hospitals (figure 5). Plant growth regulators such as Indol Butyric Acid (IBA) promote the induction of lateral roots (Sudaryono and Soleh, 1994). Kasijadi et al. (1999) reported that 75 g discarded shallot pieces added to the growing medium during air-layering propagation increased the rooting success by 10 % as compared to 7.5 ml of a 100 ppm IBA solution per plant shoot. After 5 to 6 months, the shoot has developed new roots and is ready to be separated from the parent plant (Santoso, 1990). It should be moved immediately to the field or a growth container, otherwise significant losses will occur (Santosa et al., 1996a). Usually, only 75 % of the plants from air-layering will survive. After transferred to the field, about 50 % of plants from air-layering will be damaged (Purbiati et al., 1999) due to the sensitivity to drought and strong wind (Santosa et al., 1996a). However, faster growth and better fruit quality from vegetatively propagated plants can be expected as compared to seedling plants. Young palms require heavy shading which may be reduced after one year (Schuiling and Mogea, 1992).

2.3.3. *In vitro* propagation

Salak plants can also be propagated using *in vitro* methods. Explants from generative shoots have more viable meristemoids and their adaptation capacity is better as compared with vegetative shoots (Prahardini et al., 1991). Different culture media and explant types used in *in vitro* propagation are summarised in table 1.

Table 1. Culture media and explant type recommended for *in vitro* propagation in salak

Salak type/cultivar	Explant	Medium	Effect	Reference
Salak Bali	Generative shoot	½ MS + 1 mg BA/l	Initiation phase	Prahardini et al. (1995)
	Vegetative shoot	½ MS + 10 mg IAA/l + 1 mg BA/l	Initiation phase	
	Generative and vegetative shoot	½ MS + 6 mg IAA/l	Differentiation phase	
	Generative and vegetative shoot	½ MS + 1 mg BA/l and ½ MS + 6 mg IAA + 1 mg BA/l	Multiplication phase	
	Generative and vegetative shoot	½ MS + 2 mg IAA/l + 1 mg BA/l	Shoot regeneration	
	Generative and vegetative shoot	½ MS + 6 mg IAA/l	Root multiplication	
Salak "gula pasir"	Middle part of seedling shoot	MS + 0.1 – 1.0 ppm 2,4 –D + 4.0 ppm Benzyl Amino Purin (BAP) + 200 ppm Glutamin	Shoot regeneration	Lestyana (2000)
	Middle part of seedling shoot	MS + 10 ppm IBA + 1.0 ppm BAP + 200 ppm Glutamin	Root generation	
Salak Bali	Shoot tip of seedling	½ MS + 3 mg/l BA or ½ MS + 5 mg/l Kinetin	Initiation phase	Prahardini et al. (1993)

MS : Murashige and Skoog, BA : Benzyl Adenin, IAA: Indol Acetic Acid, IBA: Indol Butiric Acid, BAP: Benzyl Amino Purin

2.4. Hybridisation

To improve quantity and quality of salak fruits, many experiments in hybridisation have been carried out. Bowo (1999) reported that the pollen fertility (male gametes) in "suwaru" exceeded 60 %. Pollen viability during storage depends however on the storage method and salak cultivar. Pollen viability of "pondoh" may be higher than 90 % when the pollen grains have been sterilised and kept for two weeks at 0 °C - 15 °C (Sumardi et al., 1995). In "kacuk", the viability of non sterilised pollen is less than 15 % after storage at –5 °C, 0 °C, 5 °C and 25 °C for 28 days, whereas the pollen viability after storage at –5 °C for 21 days was 55 % (Ashari, 1995). Three male parents, i.e. "kembangarum", "bejalen" and "suwaru" were found to increase fruit yield of salak "pondoh super" (Nandariyah et al., 2000). The hybridisation between "pondoh" and cultivars from Bali resulted in a heterosis effect concerning the Ribulose 1.5 - bis phosphate carboxylase (RuBP-carboxylase) enzyme activity and low foliar tannin content in "kelapa" x "gula pasir" and "kelapa" x "pondoh hitam" (Purnomo and Dzanuri, 1996). RuBP-carboxylase activity is closely correlated to fruit pulp thickness and sugar content, whereas leaf tannin content is correlated with fruit taste.

2.5. Flowering and Fruit Set

Salak palms will start flowering 3 to 4 years after generative propagation or after 2 to 3 years in plants derived from vegetative propagation (Schuiling and Mogea, 1992; Tjahjadi, 1993). Flowering occurs throughout the year. Female flowers open for 1 to 3 days, and more than three days without pollination will cause wilting. The best time for pollination is the second day of blossom (Tjahjadi, 1993). The male flower bunch consists of 4 to 12 inflorescences. One spike bears thousands of pollen grains. The male flower will also open for 1 to 3 days, after 3 days it will start wilting.

The natural pollination process of salak occurs by wind, rain or insects if male and female infloresences are close to each other (Tjahjadi, 1993; Bawarsiati and Rosmahani, 1994). If the distance between male and female inflorescences is too far, pollination should be done by the farmer by swaying open male flower over the receptive female inflorescence (Tjahjadi, 1993). If hand pollination is done, usually about one male inflorescence per four female inflorescences is sufficient to ensure proper pollination and fruit set. Hand pollination will often result in a better fruit quantity and quality in comparison to natural pollination (Siswandono, 1995)

The most important pollinator insect is a weevil, tentatively identified as *Nodocnemis* sp. (Schuiling and Mogea, 1992). Other insects visiting salak flowers are small black ants (*Iridomyrmex gleber* (Mayr) (Moncur and Watson, 1987), *Diptera* and *Staphilinidae* (Rosmahani et al., 1993), *Curculionidae* (Rosmahani et al., 1993; Mogea, 1978; Schuiling and Mogea, 1992), *Trigona* sp.(Hymenoptera), and *Rhynchophora palmarum* L. (Coleoptera) (Mogea, 1978). However, only *Curculionidae* seem to act as effective pollinators (Rosmahani et al., 1993; Mogea, 1978). Ants may be also active in pollen transfer (Moncur and Watson, 1987). Five beetles are necessary to pollinate one spikelet (Rosmahani et al., 1993), and ten beetles per flower bunch can be as effective as hand pollination (Baswarsiati and Rosmahani, 1994).

2.6. Maturation and Ripening

In the process of ripening, salak fruits experience an increase of weight, size, percentage of edible portion (Sosrodihardjo, 1986; Tranggono, 1998; Supriyadi et al., 2002), content of glucose, fructose sucrose, total sugar (Suhardi, 1997), sugar/acid ratio, vitamin C (Sosrodihardjo, 1986), ashes and starch (Supriyadi et al., 2002). However, a decline in the percentage of peel to fruit ratio, water content (Sosrodihardjo, 1986), tannins and acids will occur during ripening (Sosrodihardjo, 1986; Suhardi, 1995; Suhardi, 1997). Suhardi (1995) reported that content of glucose, fructose and sucrose increased until stage 6 (= 6 months after pollination) and then decreased, whereas starch content increased until stage 5.5 and thereafter declined. Supriyadi et al. (2002) reported that glucose and fructose reached the maximum content (25 % dry matter (DM)) at stage 6, however, sucrose content reached its maximum already after 5 months (16 % DM) and decreased thereafter to 10 % DM (stage 6). The contents of protein and vitamin C are reported to remain constant during maturation (Suhardi, 1995). Vitamin C content in salak fruits varies between 5 and 49 mg/100 g fresh weight (Suhardjo et al., 1995; Djaafar and Mudjisihono, 1998). In comparison to other salak cultivars (Bali, "condet" and "gading"), "pondoh" had a comparable edible portion, a higher sugar to acid ratio but lower content of tannins and acids (Sosrodihardjo, 1986). The more intensive yellow colour of fruit pulp in "pondoh" comprise of more total sugars, but less malic acid in comparison to the white pulp colour (Suhardi, 1995; Suhardi, 1997).

Supriyadi et al. (2002) reported that pulp firmness of "pondoh" increased until 174 Newton until stage 5.5, but declined to the end of ripening period (stage 6) to 130 Newton.

The major volatile compounds in salak fruit are methyl esters of butanoic acids, 2-methylbutanoic acids, hexanoic acids and the corresponding carboxylic acids (Wong and Tie, 1993; Supriyadi et al., 2002). During maturation (from stage 4.5 to 6.5), there is an increase of the esters, whereas carbonyl compounds will decrease. A significant change of volatile compounds will also occur towards ripening (Tranggono, 1998). During maturation (5 to 6 months after pollination), the methyl esters increase dramatically, exceeding the content of carboxylic acids, whereas the acid content increases reaching its maximum 6 months after pollination (Supriyadi, 2002).

Salak is a non-climacteric fruit (Suter, 1988a; Tranggono, 1998). Respiration rate in salak "pondoh" is low to moderate, i.e. 5 to 18 ml CO_2 kg^{-1} h^{-1} and declines at later maturity stages (Suhardi, 1997). Respiration rate of 5 months old salak from Bali ranged from 10 to 30 mg CO_2 kg^{-1} h^{-1}, whereas those of 6 months old fruits is 37 to 41 mg CO_2 kg^{-1} h^{-1} (Setyadjit and Sjaifullah, 1993). Respiration rate of salak fruit during storage at room temperature (29°C) after 6 days decreased from 12 - 20 mg CO_2 kg^{-1} h^{-1} to 8 - 12 mg CO_2 kg^{-1} h^{-1}. On the other hand, respiration rate of salak fruit stored at 6 °C was reduced after 4 weeks from 12 - 20 mg CO_2 kg^{-1} h^{-1} to 0.3 - 5 mg CO_2 kg^{-1} h^{-1} (Suter, 1988a).

2.7. Fruit Quality and Nutritional Valuable Compounds

Besides being a good source for antioxidants (Leong and Shui, 2002), salak is regarded to reveal a balanced content of nutritional valuable compounds (table 2).

Table 2. Nutritional valuable compounds of salak fruit (per 100 g fresh edible part)

Compound	Content
Energy (cal)	77.0[1]
Water (g)	78.0 - 82.8[1,2]
Protein (g)	1.0 – 4.0[1,2,3]
Fat (%)	0.06 – 0.08[4]
Carbohydrate (g)	20.9[1]
Sucrose (g)	10.2[2]
Glucose (g)	9.3[2]
Fructose (g)	8.1[2]
Starch (g)	4.4[2]
Pectin (g)	1.0[2]
Citric Acid (g)	0.6[2]
Malic Acid (g)	2.0[2]
Ash (g)	0.6 - 0.7[2]
Calcium (mg)	18.0 - 28.0[1,4]
Phosphorus (mg)	4.0 - 18.0[1,4]
Iron (mg)	4.2 - 7.7[2,3]
Magnesium (mg)	59.0 – 67.0[4]
Potassium (mg)	163.0 – 312.0[4]
Pro Vit A (μg RE)	170.0[5]
Vitamin B1 (mg)	0.04[1]
Vitamin C (mg)	2.0 - 2.4[1,2]
Tannin (g)	0.6[2]

Source: [1] Department of Health of Indonesia (1972)
[2] Hartanto (1998)
[3] Padmosudarso (2000)
[4] Lestari et al. (2004) (unpublished data)
[5] Setiawan et al. (2001)

2.8. Harvest

The main salak fruit production periods in West Java are from April to June and from September to October. In East Java and Bali, the main harvest periods are from January to April, whereas in Yogyakarta it is from November to January (Sudaryono et al., 1993). In order to produce a good quality, the number of fruits per bunch has to be reduced at an early fruit development stage (figure 6). Wijana et al. (1993) recommend a reduction to 10 fruits per bunch at stage 2 for cv "biasa" from Bali.

Salak fruits are usually harvested 4 to 6 months after pollination (Suhardjo et al., 1995) or at the age of 5 to 7 months (Schuilling and Mogea, 1992). The optimum harvest date seems to be 5 months after pollination (Sosrodihardjo, 1986; Roosmani and Sjaifullah, 1991; Sudaryono, 1994; Suhardjo et al., 1995). Picked at an optimal stage, "pondoh" has a shelf life of 15 days, while shelf life cultivars from Bali are only 10 days (Sudaryono, 1994). For "suwaru", it was recommended to harvest 190 to 200 days after pollination (Prabawati and Sjaifullah, 1996), whereas "condet" should be harvested 6 months after pollination (Murtiningsih and Setyadjit, 1998). Other authors recommended harvesting salak from Bali at at stage 6 (Suter, 1988a).

Fruit texture, the loss of tip peel spines, seed colour (Sosrodihardjo, 1986 and Thamrin et al., 1998), peel colour, size, aroma and flavour (Ina, 1997) are important ripening indicators of salak fruit. Rainy periods during the harvest shorten the storage life of the fruit (Thamrin et al., 1998). Salak fruits are usually picked by hand. Due to a lot of spines on the leaf mid-rib, cloves should be worn during harvest. Fruits are harvested by cutting the base of the bunch stalk with a sharp and curved knife (figure 7). Thereafter, the fruits are collected in bamboo basket with dry banana leaves as mattresses (Ina, 1997) (figure 8), to protect the fruits from physical impact. The scarce data available suggest that annual yields for salak vary from 5 to 15 t/ha (Schuiling and Mogea, 1992).

2.9. Postharvest Handling and Storage

2.9.1. Cleaning, sorting and grading

After harvest, the fruits will be immediately transported to the local markets or sold to traders, who distribute the fruits to other provinces. Cleaning is usually done by the traders with a brush before the fruits will be sold (figure 9). Also sorting (grading) is carried out by salak trader, rather than by the farmers (Ina, 1997). The purpose of this activity is to separate damaged fruits and to select the fruit according to size (grading). Grading is important to fulfill to the market demands, such as on the local markets, markets in other provinces or for export (Ina, 1997). Suter (1988a) reported that salak fruits in Bali are graded into three quality classes according to the number of fruits per kg, i.e. 9 to 10, 11 to15 and 16 to 20 fruits, respectively. On the other hand, salak "pondoh" cultivars in Yogyakarta are selected in a similar way (number of fruits per kg) into A, B and C classes, i.e. 9 to 15, 16 to 22 and 23 to 35, respectively (Departemen Pertanian, 1998). For export purposes, only class A salak fruits are suitable, for Indonesian supermarkets in big cities, class A or B are selected, whereas for local markets (figures 10, 11, 12), class B or C fruits are preferred.

2.9.2. Packaging

Due to source availability, price and mechanical injuries, the packaging of salak fruit during transport is still managed using traditional methods. This is an important tool to prevent quality losses in the postharvest process. Packaging materials used include bamboo buckets, added with chopped paper mattresses (Sulandra et al., 1987; Wijadi and Suhardjo, 1992), bamboo baskets, cardboards or wooden boxes, which are filled with banana leaves or chopped paper mattresses (Sulandra et al., 1987, Departemen Pertanian, 1998). There are three different bamboo bucket sizes used, i.e. 5 kg, 10 kg and 40 kg. Cardboards can carry 45 to 50 kg fruits, whereas wooden boxes contain 25 to 35 kg fruits (Departemen Pertanian, 1998). Packing the fruit in bunches resulted in less damage during transport than individual fruit packaging (Soedibyo and Purnomo, 1973). Damage of salak fruit during transport from Bali to Jakarta, packed in ventilated wooden boxes was much lower (0.6 % to 1.7 %) compared to those packed in bamboo baskets (50% to 60 %) (Ina, 1997). The type of package and the fruit arrangement inside the container affects significantly the percentage of fruit deterioration during transport, but not the nutritional composition (citric acid and soluble solid content) and sensory attributes (Setyadjit and Murtiningsih, 1990). Wooden boxes and bamboo baskets are appropriate for salak packaging rather than carton boxes, because they do not influence the chemical fruit characteristics during harvest and transport (Sudaryono, 1994). To protect fruits from fungal diseases during transport, fungicides (Benomyl) will be applied by dipping the fruit in a fungicide bath.

Wrasiati (1997) reported that a 10 % beewax coating of salak fruits from Bali was useful to maintain the quality, e.g. texture, content of vitamin C, tannins and organic acids and also prolonged shelf life from 7 to 12 days.

2.9.3. Storage temperature

Salak "pondoh" cultivars have superior fruit characteristics such as longer shelf life at room temperature (up to one month) (Santoso, 1990; Hastuti and Ari, 1998) as compared to other cultivars, which have a shelf-life of only for 3 to 7 days (Suter, 1988a; Santoso, 1990; Mahendra and Jones, 1993).

Cool storage will extend postharvest life of salak fruit (without information on the stage of maturity) up to 15 days, however, too low temperatures, i.e. 3 to 10 °C will cause chilling injuries (Mahendra and Janes, 1993). For salak from Bali, it was reported that at 6 °C storage temperature, shelf life of salak fruits could be extended to 28 days (Suter, 1988a). The shelf life of "pondoh" at room temperature (29 °C) was 15 days (Hastuti and Ari, 1988). At storage temperature of 15 °C, fruits maintained quality for 21 days (Djaafar and Mudjisihono, 1998), whereas at 11 °C, the storage period could be prolonged by 33 days (Hastuti and Ari, 1988).

Figure 8. Salak fruits in bamboo baskets after transport from the field

Figure 9. Cleaning of fruits with a brush

Figure 10. Fruits ready for sale on a traditional market

Figure 11. Traditional bamboo packing of fruits

Figure 12. A local market of salak fruits

During storage at room temperature, water content, starch, total sugar, organic acids of salak fruits from Bali decrease (Suter, 1988), whereas pH and softness of fruit increase (Ina, 1997). In contrast to these findings, Thamrin et al. (1998) found no changes of soluble solids content (SSC) and total acids, but a slightly reduced vitamin C content of salak "pondoh" during storage both at room temperature (27.5 °C) and at cool storage (10 °C). Whereas a panel test detected no changes in sweetness during storage under both conditions, changes in peel easiness, moisture of the fruit peel, softening and colour (brownish) of "pondoh" pulp occurred. Hastuti and Ari (1988) reported that cold storage (10 °C to 12 °C) and perforated plastic package reduced water content, tannins and sugar contents, but increased total acids in salak "pondoh". Regarding the enzyme activities, Partha et al. (1993) reported that with increasing storage temperature (4°C,

10°C, 20°C and 30°C), pectins, polygalacturonic acid and methoxyl levels of salak fruits from Bali decreased, due to the increase of activities of pectinesterase and polygalacturonase. The respiration rate of the fruit declined while the fruits gradually softened during storage.

2.9.4. Modified atmosphere packaging

The optimum storage temperature for salak in order to reduce fruit respiration is about 12 °C. For modified atmosphere packaging for salak fruits, 2 to 6 % O_2 and 10 to 18 % CO_2 should be fixed inside a low density polyethylene (LDPE) and polyvinilchloride (PVC) package with the film permeability (ß) between 0.82 and 1.9 (Purwadaria et al., 1992). The initial gas concentration of CO_2 and O_2, duration of storage and temperature regime will influence gas permeability, and concentration inside the package as well as the respiration rate. These factors are known and can be described with mathematical models (Rejo, 1996; Hartanto, 1998).

Modified atmosphere packaging in salak prolong shelf life to 35 days without influence on the sugar content (Hartanto, 1998). Amiarsi et al. (1999) reported that "lumut" fruits covered with perforated paper (0.5 cm holes), packed in 750 g PE boxes (15 x 30 cm) with initial gas concentration of 1.5 % CO_2 and 15 % O_2 and stored at 5 °C, extended shelf life to 28 days. Results of this packaging study are summarised in table 3.

Table 3. Modified atmosphere packaging of salak fruit

Salak type/cultivar	Packaging (thickness)	O_2	CO_2 %	N_2	Temperature (° C)	RH (%)	Shelf Life (days)	References
1. Salak Bali and "pondoh"	PE (0.04 mm)	15	1.5	-	15	85 - 90	24	Sudaryono (1994)
2. Salak Bali and "pondoh"	PE (0.04 mm)	15	1.5	-	10	85 - 90	15	
3. Salak Bali	PE (0.04 mm)	10	2	-	5	85 - 90	30	
4. Salak Bali	PE (0.04 mm)	15	1.5	-	15	80 - 90	24	Roosmani (1992)
5. Salak "pondoh"		3	6	91	25	-	42	Suhardi et al (1997)
6. Salak "pondoh"	PP (0.05 mm)	3	5	92	22 - 26	70 - 72	36	Lestario et al (1999)
7. Salak "pondoh"	PE (0.04 mm)	15	1.5	-	15	80 - 90	15	Roosmani (1992)
8. Salak "pondoh"	PE (30μm)	15	1.5	-	9 - 12	-	27	Roosmahani and Sjaifullah (1991)
9. Salak "nangka", "biasa", "gondok" and "nenas"	PE (80μm)	6	16	78	29	-	25	Semarajaya (1991)

PE : Polyethylene, PP : Polypropylene, RH : Relative Humidity

2.10. Processing

Salak has been processed to dry and wet candies, chips, pickles, canned fruits in syrup, juice and jam (Ina, 1997). Fresh peeled salak packed in styrofoam and covered with PE film is dipped in 300 ppm ascorbic acid and 0.05 % sodium benzoate and stored at 10 °C for approximately 8 days (Prabawati et al., 1994). Fruits of the same treatment stored at 0 °C can be kept in good condition for 20 days (Amiarsi et al., 1999). The application of edible film coating (based on metoxy pectin with 0.25 % stearic acid) will prolong shelf life 3 days to 5 days for "pondoh" (Wuryani, 1999). This edible film coating maintains moisture content, total sugars, ascorbic acid concentration, total acidity, tannin content and textural properties at 5 °C, but not at 10 °C (Setiasih, 1999). For salak chips, the optimal maturity stage is 5.5 to 6 months for salak from Bali and "pondoh", respectively, at drying temperatures of 70 to 80 °C (Wijadi and Suhardjo, 1994). To maintain the aroma of salak, juice products should be prepared within 3 days after harvest. Mudjisihono and Handayani (2000) reported that the addition of 0.5 % to 1 % carboxyl methyl cellulose increased the viscosity of the juice and the suspension stability of the juice for one week. Thamrin et al. (1998) reported that "pondoh super" juice had relatively less sediment as compared with "gading", "pondoh hitam" and "pondoh manggala" cultivars. Sunarmani (1988) reported that for preparing pickles in syrup, dipping salak fruits in sodiumbisulfide (75 ppm) for about 1 h prevented from enzymatic browning. The browning of dried salak fruit was inhibited by adding 1.5 g/l sodiumbisulfide and 1 g/l citric and ascorbic acid (Prabawati et al., 1994). Ina (1997) reported that dipping in concentrated sugar and sodiumbisulfide solution (1200 ppm) retarded browning of candied fruit segments. Dipping the fruit in bisulfide solution (1 %, 1 h) before canning effectively protects the fruit from browning.

2. 11. Pest and Diseases

The main pests in salak "pondoh" are beetles such as *Omotemnus serrirostris* (*Curculionidae*) (figure 13), *Omotemnus miniatocrinitus* (figure 14) (Mangoendihardjo, 1975; Tjahjadi, 1989) and *Silphidae* (Mahfud et al., 1994). *Pseudococcus* sp. and several mammals species mainly attack salak from Bali (Mahfud et al., 1993). *Silphidae* insects enter the fruit from its top. The whole life cycle of this insect from egg to adult is only 33 days (Mahfud et al., 1994). Weevils (*Nodocnemis* spp.), which bore into the young fruit, are occasionally harmfull (Schuilling and Mogea (1992). Other pests that feed on salak palms include the monophagous beetle *Calispa elegans*, the polyphagous caterpillar *Ploneta diducta*, a leaf roller *Hidari* sp. (figure 15), the scale *Ischnapsis longirostris* and the bug *Tolumnia* sp. (figure 16) as well as rodents such as rat and "luwak" (*Paradoxurus hermaphroditus*) (Schuilling and Mogea, 1992). These rodents usually eat and destroy salak fruits (figure 17).

The most important diseases on salak are: leaf spot (*Pestaliopsis palmarum*), wilting of flowers (*Fusarium* sp. and *Marasmices polmivorus*), fruit wilt (*Ceratocystis paradoxa*, *Fusarium* sp. and *Aspergillus* sp.) and fruit malformation (Mahfud et al. 1993). Schuiling and Mogea (1992) reported on a fungal disease developing a white mycelium (*Mycena* sp.) and on a "pink disease" (*Corticium salmonicolor*). Salak cultivars from Bali and "pondoh" are found to be susceptible to leaf spot (Mahfud et al., 1994). A fungus species attacking fruit after harvest or during storage and marketing is *Thielaviopsis* sp. Cracking of fruit peel occurs only less than 1 %, especially in the rainy season (figure 18) (Ina, 1997).

Mahfud et al. (1993) reported that a combination of manual control and systemic fungicide application via roots was the most effective measure to control *Silphidae*, flower and fruit wilt caused by *Ceratocystis paradoxa*, *Fusarium* sp. and *Aspergillus* sp. However, manual control by cutting the infected plant parts or by manual searching and killing the insects were preferred due to the problem of pesticide residues in fruits (Tjahjadi, 1993; Mahfud et al., 1993). Hot water treatment (60 °C for 30 min) was the most effective method to control postharvest fungus diseases. Hot water dipping of fruits at 50 °C for 3 min retarded fruit rot caused by *Thielaviopsis* sp. for 16 days storage period at 15 °C or 10 days at room temperature (29 °C) (Murtiningsih et al., 1995).

6.5 cm

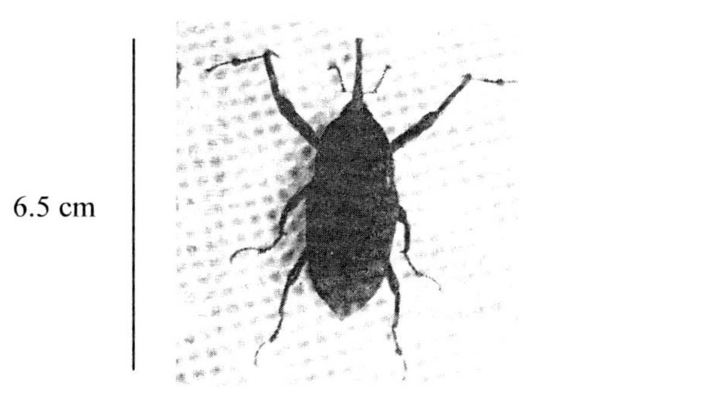

Figure 13. *Omotemnus serrirostris*

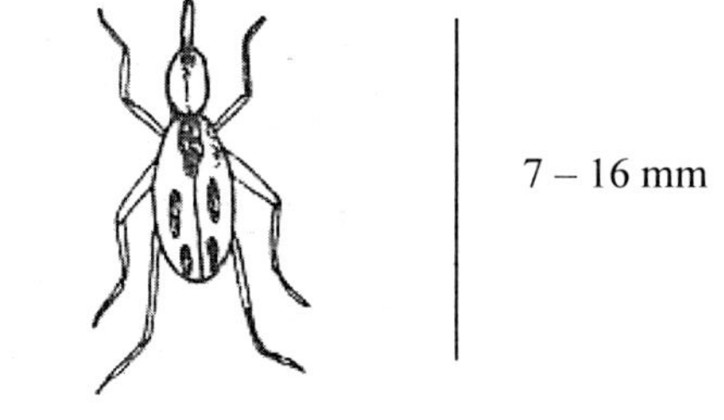

7 – 16 mm

Figure 14. *Omotemnus miniatrocnitus*

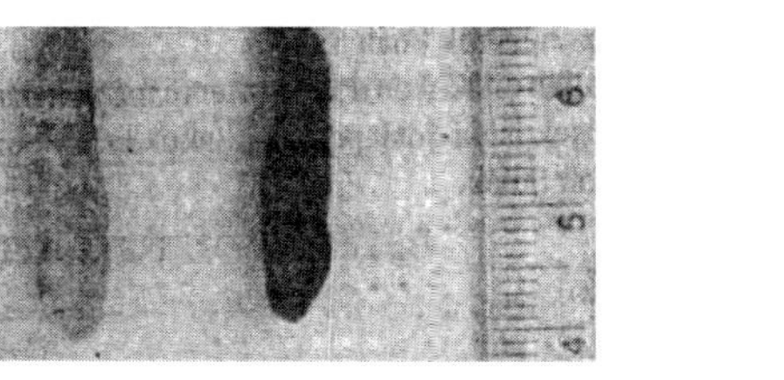

Figure 15. *Hidari* sp.

Source: Tjahjadi (1989)

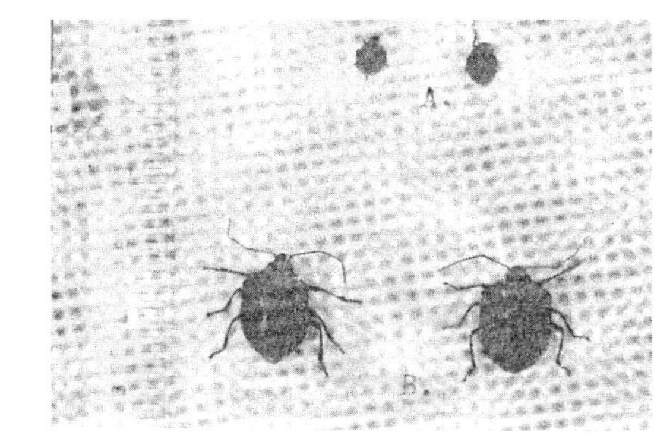

Figure 16. *Tolumnia* sp.

Figure 17. Salak fruits eaten by rodents

Figure 18. Cracking of salak fruits

3. RESPONSES OF SALAK CULTIVARS (*Salacca zalacca* (Gaertn.) Voss) TO DIFFERENT GROWING MEDIA

3.1. Introduction

Salak belongs to a group of palms, which are 1.5 - 5 m high, extremely spiny and sprout their leaves from the ground level. Salak palms grow as under-storey plants in the low lands of tropical rain forests in Indonesia and other Southeast Asian countries. In Indonesia, salak has been cultivated throughout the islands and the fruit is widely used as fresh fruit. One important salak cultivar is "pondoh" due to its superior fruit quality. "Pondoh" is cultivated in the sub district Sleman of the Yogyakarta province. "Gading" is another important cultivar grown in the same area as "pondoh".

Indonesia has many options for the development of the horticultural industry, such as intensive plantation systems, year-round production and high quality of fruits (Winarno, 1997). Therefore, the development of the horticultural sector including fruit production has priorities such as in choosing suitable commodities for certain areas and the promotion of fruit production. Salak production is also included in the priority list. New salak plantations have been made from 1991 in Riau (Sumatera island) and in Kalimantan (Winarno, 1997). In the new production areas, suitable site conditions for salak palms such as peat-land have to be found without neglecting the quality of the product.

In Indonesia, wetlands cover an area of 17 to 27 million ha, which are distributed over Sumatera, Kalimantan, Sulawesi and Irian Jaya. Only 5.6 million ha can be used for agricultural purposes (Subagyo et al., 1996). However, the development of the agricultural sector is very fast. Therefore, the possibility of using peat land for horticultural purposes such as for salak is an interesting point. A study showed that a mixture of peat/rice hull (2/1, v/v) was suitable as growing media for salak "pondoh" seedlings (Kusmiba, 1994). In industrialised countries, peat alone or mixed with other organic and inorganic substances, is one of the most important materials for the preparation of pot plant substrates (Cattivello et al., 1997).

Peat is a mixture of partially decomposed plants, microbes and their transformation products, which grow in swampy conditions. The water holding ability of peat is high, but aeration in containers may be poor (Handreck and Black, 1999). Tropical peat soil has a low mineral content, a low pH (< 4) but about 90 % of organic matter (Andriesse, 1988). The content of nitrogen is very low due to the high C/N ratio. Micro-nutrients such as Cu, B and Zn are also leaching in organic soils (Subagyo et al., 1996).

Compost is another growing media usually used in horticulture, which results from controlled decomposition of organic matter to a point where the product can be beneficially used to improve crop productivity. The low concentration of nutrients in compost, particularly nitrogen and potassium, causes a reduced plant growth. Therefore, fertilisers are added to compost (Chaney et al., 1980). High organic matter content, porosity, good aeration and drainage characteristics are some of the benefits of compost (Pinamonti and Sicher, 2000). However, some negative aspects of compost such as physiological disorders of crop plants, eutrophication of the ground water and nutrient losses from the substrate should also be considered (Pinamonti and Sicher, 2000).

Sand culture in combination with the use of nutrient solutions has been used in horticultural production since the 19th century (Hewitt, 1952). Quartz sand, which is low in calcium carbonate (limestone) and having a grain size of about 1.2 mm in diameter, allows good aeration of plant roots and it retains sufficient moisture (Ellis and Swaney, 1938)

In Indonesia, salak production techniques and postharvest properties of the fruit have been investigated during the last years. To promote the production and the fruit quality of salak, knowledge about ecophysiological properties is very important. However, only a few investigations of these aspects have been conducted so far. The purpose of this study was to investigate growth and physiological responses of different salak cultivars to different plant growing media.

3.2. Materials and Methods

3.2.1. Location and experimental design

The study was carried out from November 2002 until February 2003 for a period of 14 weeks in the greenhouse of the Fruit Science Department in Berlin-Dahlem. The temperature inside the greenhouse were adjusted to 20 °C/25 °C (day/night) and the relative humidity varied in between 20 % and 70 %. Two 400 W lamps (HQI-TS/D, OSRAM, Germany) at about 2 m above the salak seedlings were set up for 12 h (6 am to 6 pm) to provide additional light.

The experiment (a total of 84 plants) was arranged in block design. Four-months-old seedlings of 4 salak cultivars were used for the study, i.e. "gading" (G), "pondoh super" (PS), "pondoh hitam" (PH) and "pondoh manggala" (PM). Three types of plant growing media were used: 1. sand (S), 2. peat (P) and 3. sand/compost mixture (SC; 1/1, weight/weight). Media and cultivar were tested in 12 combinations: SG, SPS, SPH, SPM, PG, PPS, PPH, PPM, SCG, SCPS, SCPH and SCPM. Pure quartz sand (0.6 to 1.2 mm) was used as control media. Peat medium consisted of 95 % to 98 % organic matter with H3 – H5 degree of decomposition and pH 3. The compost medium (pH 6) comprised of 320 g/l N, 320 g/l P_2O_5, 450 mg/l K_2O and < 3 g/l NaCl. After germination, seedlings were transplanted in 14.5 cm x 11 cm pots, filled with growing media according to the treatments listed above and were watered daily (100 ml/plant) with a complete nutrition solution (Manna Lina Spezial 24–5–11 (3), Wilhelm Hang, Germany), prepared with distilled water. The nutrition solution contained 24 % N, 5 % P_2O_5, 11 % K_2O, 3 % MgO, 0.05 % Fe, 0.05 % Mn, 0.005 % Zn, 0.02 % Cu, 0.01 % B and 0.002 % Mo.

The following parameters were measured: shoot dry weight, root dry weight, shoot length increment, leaf area increment, net CO_2 assimilation rate (P_N), leaf colour and plant mineral contents (N, P, K, Ca and Mg).

3.2.2. Growth

Shoot and root dry weight were measured at the end of the study after drying the fresh samples at 103 °C to constant weight (Maier, 1990). Leaf samples were weighed after drying at 65 °C for 8 h and were later used for the analysis of mineral contents. Shoot length and leaf area were first measured immediately after the seedling had been transplanted into the growing media. The second measurement was 14 weeks later. Shoot length was measured with a ruler, leaf area was measured with a leaf area meter (CI – 202, CID Inc., USA).

3.2.3. Net CO_2 assimilation rate

Net CO_2 assimilation rate (P_N) of seedling leaves was assessed 6, 8 and 13 weeks after the onset of the experiment using a portable photosynthesis system (CI-301PS, CID Inc., USA). A 400 W lamp (HQI-TS/D, OSRAM, Germany) was used to provide additional light during the gas exchange measurements (starting from 10 am until 1 pm). Before the measurement, plants were kept under the lamp for about 1 h for adaptation.

3.2.4. Leaf colour

The measurement of leaf colour was conducted using a Minolta Colorimeter (CR-321, Minolta, Japan) with a standardised light type D65. The equipment was calibrated with a white standard plate. Colour measurement were expressed in the L* a* b* scale, where L* indicates the luminiscence on a 0 to 100 scale from black to white. The colour coordinates of a* and b* locate on a rectangular-coordinate grid perpendicular to the L* axis with 0 to 60 scale. The colour at the grid origin (a* = 0, b* = 0) is achromatic (gray). On the horizontal axis, positive a* indicates red and negative a* green colour. On the vertical axis, positive b* indicates yellow and negative b* blue colour. Chroma represents the hypotenuse of a right triangle created by joining points (0, 0), (a*, b*) and (a*, 0). As chroma increase, colour becomes more intense. Hue angle is the angle between the hypotenuse and 0° on the a* (green/red) axis. Hue angle of 0° indices red, 90° yellow, 180° green and 270° blue.

Chroma was calculated as follows:

chroma $= \sqrt{(a^*)^2 + (b^*)^2}$,

whereas hue angle was computed as follows:

hue angle = $\tan^{-1}$ (b*/a*) (McGuire, 1992).

3.2.5. Mineral content

Leaf mineral contents were analysed from dried and ground samples (Mikro-Feinmühle-Cullati DCFH 48, Janke & Kunkel, Germany). Nitrogen content was determined following a modified Kjeldahl method as has been described by Okoye (1980). 500 mg leaf powder were digested (Büchi 430 Digestor, Büchi Labortechnik AG, Switzerland) in boiling 98 % H_2SO_4 (20 ml) and selenium catalyst (15348 Merck, Germany). The samples were thereafter distilled in a Büchi-Scrubber B-412 (Büchi Labortechnik AG, Switzerland) into 0.1 mol H_3BO_4 and subsequently were titrated with 0.1 N HCl. Nitrogen content was calculated and expressed as % of DM.

P, Ca, Mg and K were analysed according to a modification method as described by Evenhuis and de Waard (1980). The samples were dry-ashed at 490 °C for 4 h. Ashes were dissolved in 25% HCl and were evaporated in a sand bath for 20 min until the complete disappearance of solid residues. P was determined photometrically (Eppendorf 1101 M, Germany). The other minerals were analysed using atomic absorption spectophotometry (AAS) (905 A, GBC, Australia). The data of the minerals were expressed as mg/g DM.

3.2.6. Statistical analyses

All sample data were subjected to compare means and further analysed with least significant difference (LSD) test (Steel et al., 1996), with significant difference of $P \leq 0.05$ between means using the Statistic Program MINITAB 14 (Minitab Inc., State College, USA, 2001). The analyses comprised of solely the effect of salak cultivars or growing media to growth, P_N, leaf colour and mineral content of salak seedlings.

The effect of interactions between salak cultivars and growing media on growth, P_N, leaf colour and mineral content were also analysed.

3.3. Results and Discussions

3.3.1. Growth

The seedling age in this study (four months after germination) is representative for the establishing conditions in the nursery before the plants are moved to the field. Growth parameters of different salak cultivars are presented in figure 19 and table 4. Among the different salak cultivars, only small differences in the growth parameters were found. "Pondoh hitam" showed a stronger shoot growth than the other cultivars (19 % to 48 %). Leaf growth increments of "pondoh hitam" and "pondoh manggala" cultivars were higher as compared with that of "gading" and "pondoh super".

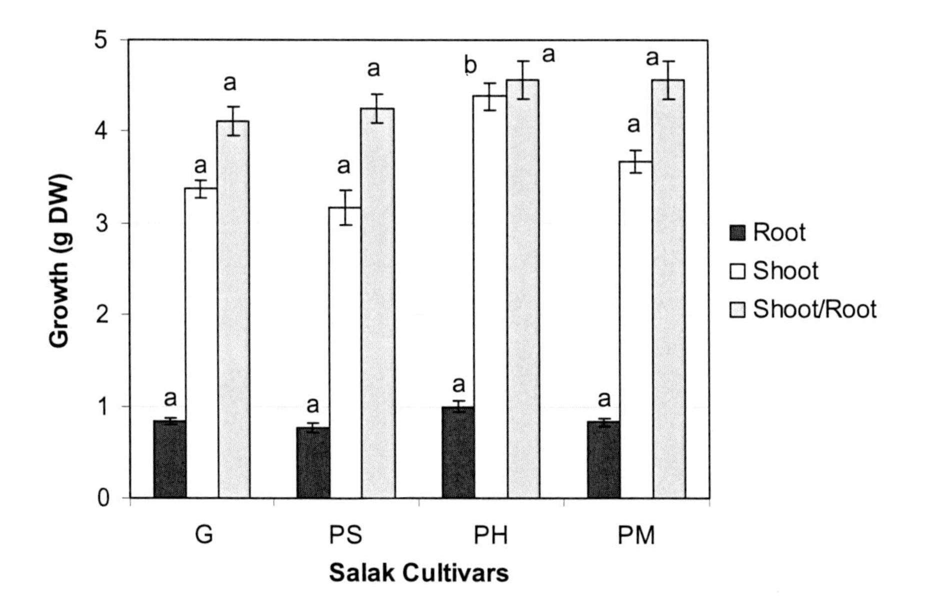

Figure 19. Root and shoot dry weight and shoot/root ratio of different salak cultivars ("gading" (G), "pondoh super" (PS), "pondoh hitam" (PH) and "pondoh manggala" (PM)). Different letters in the same parameter indicate significant differences by LSD test ($P \leq 0.05$)

Table 4. Leaf area increment (LAI) and shoot length increment (SLI) of different salak cultivars ("gading" (G), "pondoh super" (PS), "pondoh hitam" (PH) and "pondoh manggala" (PM))

Growth parameter	Cultivar			
	G	PS	PH	PM
LAI (cm^2)	213.3 ± 17.4 a	240.8 ± 7.9 a	288.2 ± 15.7 b	280.0 ± 22.3 ab
SLI (cm)	44.98 ± 1.19 a	41.09 ± 2.21a	49.89 ± 2.28 a	43.38 ± 3.01 a

Values represent means ± SE. Different letters in the same row indicate significant differences by LSD test ($P \leq 0.05$)

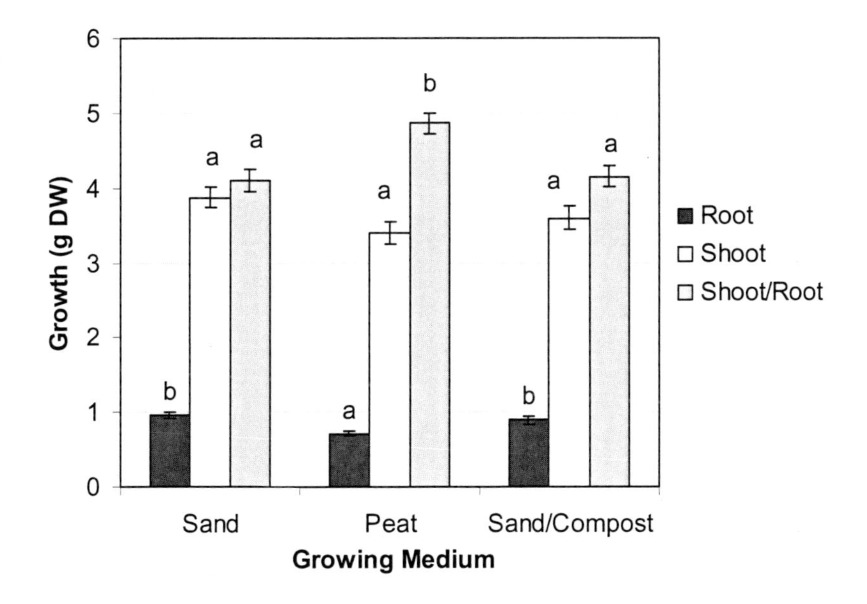

Figure 20. Root and shoot dry weight and shoot/root ratio salak seedlings growing in different media. Different letters in the same parameter indicate significant differences by LSD test ($P \leq 0.05$)

Table 5. Leaf area increment (LAI) and shoot length increment (SLI) of salak seedlings growing in different media

Growth parameter	Growing medium		
	Sand	Peat	Sand/Compost
LAI (cm^2)	300.5 $\pm$ 10.7 b	264.9 $\pm$ 10.2 b	187.0 $\pm$ 15.3 a
SLI (cm)	50.48 $\pm$ 1.46 b	44.46 $\pm$ 1.86 ab	38.55 $\pm$ 1.99 a

Values represent means $\pm$ SE. Different letters in the same row indicate significant differences by LSD test ($P \leq 0.05$)

Figure 21. Root growth of salak "pondoh manggala" (PM) seedlings in different growing media (peat (P), sand (S), sand/compost (SC))

The influence of different media on growth parameters of salak seedlings are summarised in figure 20 and table 5. Seedlings growing on peat had 27 % and 21% less root dry weight than those growing in sand or in sand/compost, respectively (figure 20 and 21). The shoot/root ratio of peat-grown seedlings was higher as compared to that of the other variants. On the other hand, leaf area and shoot length increment of salak growing in sand/compost was lower as in seedlings grown in sand or in peat.

Table 6. Effect of salak cultivars ("gading" (G), "pondoh super" (PS), "pondoh hitam" (PH), "pondoh manggala" (PM)) and growing media on growth of salak seedlings

Growth parameter	Cultivar	Growing Medium		
		Sand	Peat	Sand/Compost
Root DW (g)	G	0.94 ± 0.07 c	0.81 ± 0.04 bc	0.77 ± 0.06 bc
	PS	0.86 ± 0.06 bc	0.54 ± 0.02 a	0.93 ± 0.13 bc
	PH	1.11 ± 0.11 c	0.82 ± 0.13 abc	1.00 ± 0.09 c
	PM	0.96 ± 0.07 c	0.70 ± 0.04 b	0.83 ± 0.10 bc
Shoot DW (g)	G	3.56 ± 0.22 bc	3.43 ± 0.15 bc	3.14 ± 0.08 b
	PS	3.26 ± 0.17 bc	2.55 ± 0.11 a	3.79 ± 0.49 bcd
	PH	4.68 ± 0.20 d	4.28 ± 0.43 cd	4.15 ± 0.22 cd
	PM	3.97 ± 0.23 cd	3.69 ± 0.08 c	3.12 ± 0.17 b
Shoot/Root	G	3.94 ± 0.39 ab	4.25 ± 0.16 ab	4.16 ± 0.22 ab
	PS	3.83 ± 0.23 a	4.74 ± 0.23 b	4.16 ± 0.25 ab
	PH	4.39 ± 0.25 ab	5.37 ± 0.38 b	4.28 ± 0.33 ab
	PM	4.23 ± 0.27 ab	5.31 ± 0.22 b	3.85 ± 0.34 ab
LAI (cm^2)	G	272.9 ± 17.7 c	242.2 ± 15.1 bc	124.9 ± 20.9 a
	PS	259.9 ± 12.1 bc	219.9 ± 12.8 b	243.0 ± 12.7 bc
	PH	325.4 ± 13.9 c	324.0 ± 15.3 c	230.5 ± 25.2 bc
	PM	343.8 ± 24.1 c	298.9 ± 8.69 c	135.3 ± 28.5 a
SLI (cm)	G	45.70 ± 2.50 cd	45.86 ± 1.18 c	43.37 ± 2.42 c
	PS	49.11 ± 2.12 cd	37.34 ± 3.46 bc	36.10 ± 3.95 b
	PH	55.34 ± 2.55 d	51.85 ± 4.32 bcd	43.33 ± 3.69 a
	PM	51.74 ± 3.55 cd	45.97 ± 4.04 bcd	28.04 ± 2.96 ab

DW= Dry weight, LAI= Leaf area increment, SLI= Shoot length increment

Values represent means ± SE. Different letters in the same parameter indicate significant differences by LSD test ($P \leq 0.05$)

The effect of cultivar and growing media on the growth of salak seedlings are presented in table 6. There was a slight reduction in root growth of salak cultivars growing in peat (figure 21). On the other hand, "pondoh hitam" was found to have the most vigorous root and shoot growth in all growing media tested. "Pondoh super" seedlings growing in peat were found to have the lowest root and shoot dry weight. The tendency of higher shoot/root ratio was found in all salak cultivars growing in peat. Leaf area and shoot length increment of "pondoh hitam" were higher in all media tested as compared with the other cultivars. "Pondoh manggala" grown in sand/compost was found to develop the lowest leaf area and shoot length increments.

Growth data indicated that "pondoh hitam" seems to be the most vigorous cultivar, since the growing conditions were same for all cultivars tested. Many experimental studies revealed that plant show genetically controlled differences in dry matter production (Grime and Campbell, 1991). On the other hand, the results of the effect of growing media on growth showed that sand was the best medium for salak seedlings. Sand enables a good aeration for the roots and a high availability of nutrients. The low pH of peat induced a decrease of root growth and an increase in shoot/root ratio in salak plants. Inhibition of root elongation at pH values below 5 is a common feature in many plant species and is caused by various factors, such as impairment of H^+ efflux from roots (Schubert et al., 1990). According to Marschner (1995), pH effects on cation uptake are in accordance with the key role of the plasma membrane-bound proton efflux pump as the driving force for ion uptake. At high substrate H^+ concentration, the efficiency of the H^+ efflux pump decreases whereas the downhill transport of H^+ from the apoplasm into the cytoplasm is enhanced. Electropotential of root cells decreases from –150mV at pH 6 to –100mV at pH 4 (Dunlop and Bowling, 1978). On the other hand, some possible changes in physical and chemical characteristics during aging of compost media could reduce leaf area and shoot length of salak seedlings. Some composts in pots have been reported to have satisfactory pore space at the beginning of the period, but undergo compaction during the further growth period of the plants (Fitzpatrick, 2002). Moreover, according to Handreck and Black (1999), high moisture conditions of composts supplemented with fertiliser will cause continuous decomposition. The compost may become anaerobic at low pH or harmful compounds such as organic acids will accumulate. The results of this study concerning interaction of cultivar and growing media indicated that "pondoh hitam" was tolerant to all growing media tested. On the other hand, higher susceptibility of "pondoh super" to peat rather than the other media tested was found. Cattivello et al. (1997) reported that different kind of peat media and plant species significantly influenced growth parameters and the final quality of *Cyclamen persicum* and *Euphorbia pulcherrima*.

3.3.2. Net CO_2 assimilation rate

The net CO_2 assimilation rate (P_N) of different salak cultivars is presented in figure 22. P_N of all cultivars increased from week 6 to week 13 after transplanting. "Gading" exhibited the highest P_N of all cultivars tested. P_N of "gading" at week 6 was up to 100 % higher, whereas at week 13 was up to 35 % higher than P_N of the other cultivars.

The effect of growing media on P_N of seedlings is summarised in figure 23. P_N of salak seedlings growing in peat and in sand/compost was higher than that of seedlings in sand, except at week 13 after transplanting.

The interactive effect of salak cultivars and growing media on P_N is presented in table 7. P_N results reflected a higher P_N of "gading" in all growing media tested as compared with the other cultivars. It seems that the genotype has a stronger effect on P_N than the growth substrate especially at an early developmental stage. This result is consistent with findings of Salisbury and Ross (1985). Differences in P_N of the salak cultivars tested in this study can be explained by the fact that "gading" is not closely related to the pondoh cultivars.

P_N of salak plants is relatively low (0.7 to 4.5 $\mu mol\ CO_2\ m^{-2}\ s^{-1}$) as compared to other tropical tree species, such as peach palm, *Bactris gasipaes* Kunth (10.1 $\mu mol\ CO_2\ m^{-2}\ s^{-1}$) (Oliveira et al., 2002), oil palm (25 $\mu mol\ CO_2\ m^{-2}\ s^{-1}$) (Dufrene and Saugier, 1993), mango (5.9 to 9.6 $\mu mol\ CO_2\ m^{-2}\ s^{-1}$) (Fukamachi et al., 198) and banana (15 to 22 $\mu mol\ CO_2\ m^{-2}\ s^{-1}$) (Thomas and Turner, 2001).

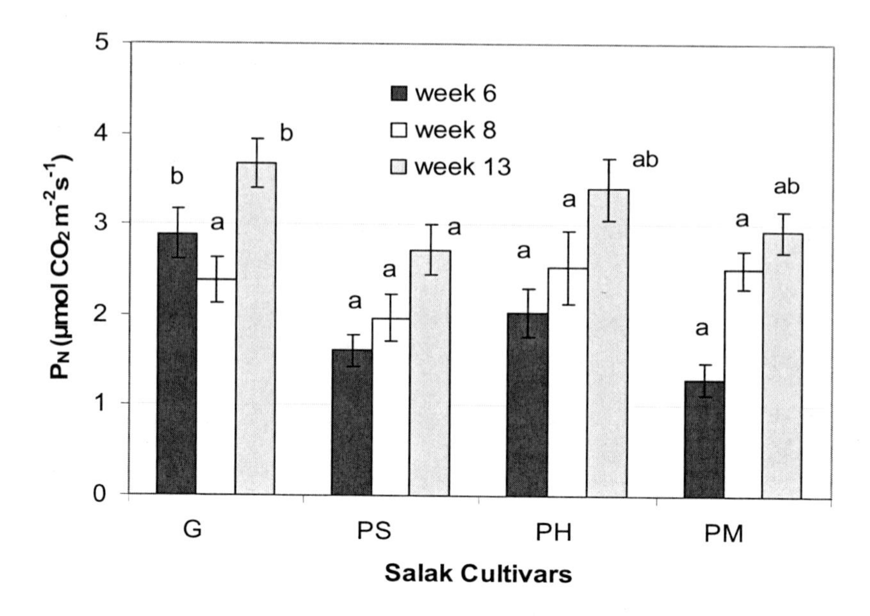

Figure 22. Net CO_2 assimilation rate (P_N) of different salak cultivars, "gading" (G), "pondoh super" (PS), "pondoh hitam" (PH) and "pondoh manggala" (PM) at week 6, 8 and 13 after transplanting. Different letters in the same parameter indicate significant differences by LSD test ($P \leq 0.05$)

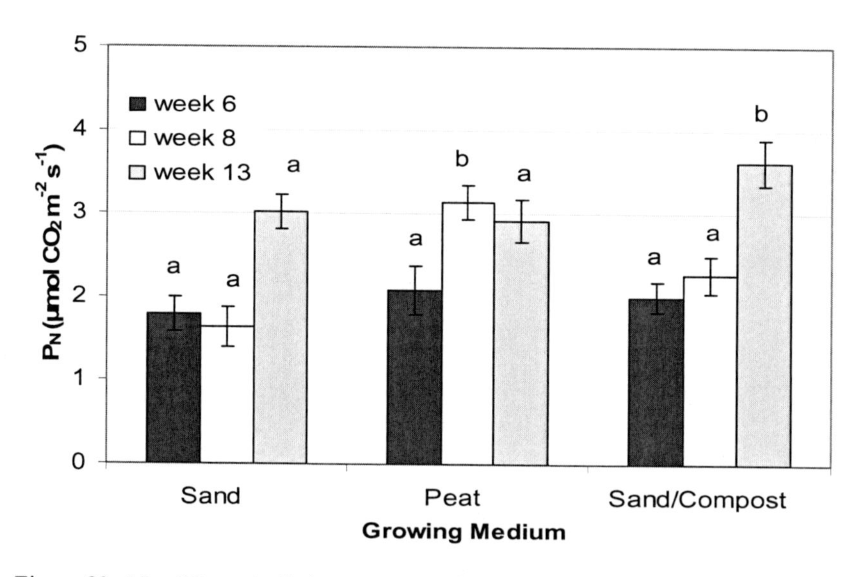

Figure 23. Net CO_2 assimilation rate (P_N) of salak seedlings growing in different media at week 6, 8 and 13 after transplanting. Different letters in the same parameter indicate significant differences by LSD test ($P \leq 0.05$)

Table 7. Effect of salak cultivars ("gading" (G), "pondoh super" (PS), "pondoh hitam" (PH), "pondoh manggala" (PM)) and growing media on net CO_2 assimilation rate (P_N) of salak seedlings on week 6, 8 and 13 after transplanting

P_N ($\mu mol\ CO_2\ m^{-2}\ s^{-1}$)	Cultivar	Growing Medium		
		Sand	Peat	Sand/Compost
Week 6	G	2.95 ± 0.11 c	3.60 ± 0.32 c	2.13 ± 0.65 abc
	PS	1.50 ± 0.25 ab	1.40 ± 0.38 ab	1.92 ± 0.26 b
	PH	1.33 ± 0.47 ab	2.57 ± 0.53 bc	2.20 ± 0.28 bc
	PM	1.37 ± 0.37 ab	0.75 ± 0.19 a	1.77 ± 0.20 b
Week 8	G	2.27 ± 0.47 ab	3.15 ± 0.28 b	1.75 ± 0.38 ab
	PS	1.52 ± 0.30 ab	3.05 ± 0.43 bc	1.34 ± 0.28 ab
	PH	0.73 ± 0.52 a	4.12 ± 0.21 c	2.77 ± 0.42 b
	PM	2.05 ± 0.41 ab	2.23 ± 0.27 ab	3.25 ± 0.21 b
Week 13	G	2.88 ± 0.29 ab	3.75 ± 0.46 abc	4.38 ± 0.48 b
	PS	2.16 ± 0.37 a	2.55 ± 0.42 a	3.47 ± 0.55 abc
	PH	3.60 ± 0.41 abc	2.12 ± 0.63 a	4.48 ± 0.20 c
	PM	3.42 ± 0.35 abc	3.25 ± 0.40 ab	2.13 ± 0.18 a

Values represent means ± SE. Different letters in the same parameter data indicate significant differences by LSD test ($P \leq 0.05$)

3.3.3. Leaf colour

Leaf colour can be used to determine the nutritional state of a plant (i.e. nutrient deficiency) (Marschner, 1995). This parameter provides also valuable information on the mineral content of leaves. Data on leaf colour of salak seedlings of different cultivars are presented in table 8. No differences were found for L*, a*, b*, chroma and hue angle values between the cultivars tested, indicating there was no cultivar effect on the leaf colour of salak seedlings.

Data on leaf colour of seedlings grown in different growing media are summarised in table 9. Foliar L*, b* and chroma values of seedlings growing in peat were significantly higher than of plants of the other variants, whereas a* and hue angle values of seedlings growing in peat were lower as compared to sand and sand/compost. The data reflected also a darker green leaf colour of seedling grown in sand or sand/compost than in peat.

In contrast, these results point out a more pronounced yellow leaf colour of seedlings growing in peat, which was possibly caused by nutrient imbalances.

Table 8. Colour parameters (L*, a*, b*, chroma and hue angle) in leaves of different salak cultivars ("gading", "pondoh super", "pondoh hitam" and "pondoh manggala")

Colour parameters	Cultivar			
	"gading"	"pondoh super"	"pondoh hitam"	"pondoh manggala"
L*- value	37.41 ± 0.93 a*	37.81 ± 0.70 a	36.97 ± 0.63 a	37.90 ± 0.58 a
a* - value	-13.73 ± 0.52 a	-14.24 ± 0.53 a	-13.15 ± 0.54 a	-14.46 ± 0.39 a
b*- value	17.22 ± 1.23 a	18.01 ± 1.12 a	16.37 ± 0.98 a	18.33 ± 0.69 a
Chroma	22.05 ± 1.29 a	22.98 + 1.20 a	21.01 + 1.11 a	23.36 + 0.76 a
Hue angle	129.11 ± 0.87 a	128.70 + 0.70 a	129.11 + 0.57 a	128.41 + 0.58 a

Values represent means ± SE. Different letters in the same row indicate significant differences by LSD test ($P \leq 0.05$)

Table 9. Colour parameters (L*, a*, b*, chroma and hue angle) in leaves of salak seedlings growing in different media

Colour parameters	Growing medium		
	Sand	Peat	Sand/Compost
L*- value	35.73 ± 0.33 a*	39.73 ± 0.60 b	37.11 ± 0.40 a
a*- value	-13.05 ± 0.35 b	-15.26 ± 0.37 a	-13.37 ± 0.40 b
b*- value	15.69 ± 0.60 a	20.47 ± 0.82 b	16.28 ± 0.70 a
Chroma	20.42 ± 0.68 a	25.55 ± 0.87 b	21.08 ± 0.78 a
Hue angle	129.93 ± 0.38 b	126.95 ± 0.54 a	129.61 ± 0.51 b

Values represent means ± SE. Different letters in the same row indicate significant differences by LSD test ($P \leq 0.05$)

3.3.4. Mineral content

The influence of growing media on mineral contents of salak seedlings is summarised in table 10. Salak leaf mineral concentration was similar to those found in 4-months-old oil palm seedlings, i.e. 22 to 32 mg/g N, 1.4 to 2.1 mg/g P, 12 to 15 mg/g K, 3 to 5 mg/g Mg 5 to 6 mg/g Mg (Ross, 1999). Mineral contents in leaves of most salak seedlings growing in different media tested were in the optimal range, due to the supply of a balanced nutrition solution to the plants. Higher contents of Mg and P of salak seedlings in peat in comparison with that in other growing media were found. Low content of P of seedlings in sand/compost was observed. N, P and Mg contents of seedlings in sand/compost were much lower, but K and Ca were higher than those of seedlings growing in the other media.

Table 10. Mineral contents of salak seedlings growing in different media

Growing media	N (mg/g)	P (mg/g)	K (mg/g)	Mg (mg/g)	Ca (mg/g)	Σ
Sand	44.18 ± 1.7 b	3.69 ± 0.1 b	8.96 ± 0.3 a	1.88 ± 0.1 a	6.10 ± 0.2 a	64.81
Peat	44.76 ± 1.0 b	6.86 ± 0.2 c	11.83 ± 0.2 b	3.70 ± 0.5 b	7.65 ± 0.3 ab	74.80
Sand/ Compost	32.45 ± 1.6 a	2.48 ± 0.1 a	14.37 ± 0.6 c	1.44 ± 0.1 a	9.48 ± 0.7 b	60.22

Values represent means ± SE. Different letters in the same column indicate significant differences by LSD test ($P \leq 0.05$)

Note:

Optimal range for plant growth according to Marschner (1995) as follows (in mg/g):

N = 20 – 50 P = 3 -5 K = <20 – 50
Mg = 1.5 – 3.5 Ca = 1 - > 50

Optimal leaf element concentration for oil palm (*Elaeis guinensis*) according to Fairhorst and Härdter (2003) as follows (in mg/g):

N = 14 P = 1.5 – 1.9 K = 10 – 13
Mg = 1.6 – 2.2 Ca = 1.4 – 2.5

Total amounts of nutrients in the solution applied for each salak seedling of all treatments during the study (14 weeks) were 260.6 mg N, 52.5 mg P, 22.3 mg K and 44.2 mg Mg. No Ca was applied to the plants. There was a considerable variation amount of nutrient levels in leaves of salak seedlings growing in different media (table 10). These indicated a shift in nutrients uptake due to the medium used. Marschner (1995) explained that the concentration of nutrient in the soil (media) solution is a primary importance for nutrient supply to root. The concentration varies widely depending on such factors as moisture, pH, cation-exchange capacity, redox potential, quantity of organic matter, microbial activity and fertiliser application (Asher, 1978). The higher Ca content in leaves of seedlings growing in peat and sand/compost than in sand indicated the availability of Ca in peat and compost media, since no additional supply of Ca to salak plants during the study. Moreover, the availability of Mg and P in peat supplied with additional fertiliser was high, leading to luxurious content of these mineral in salak seedlings. In contrast, N, P and Mg might have been fixed in the compost media, and became unavailable for plants due to the factors mentioned above (moisture, pH, cation-exchange capacity, redox potential, quantity of organic matter, microbial activity and fertiliser application). Moreover, the competition of cations in compost media can be seen as the reason for the different mineral content results of seedlings in sand/compost. Soil organic matter can fix considerable amounts of ammonia in non-exchangeable forms (Marschner, 1995). The competition among cations was possibly enhanced by the application of minerals, for example between K^+ and NH_4^+ (Axley and Legg, 1960; Walsh and Murdoch, 1963). Competition in ion uptake by salak seedlings in growing media tested could also be a reason for the diverging of nutrient contents of the plants. This competition occurs particularly between ions with similar physicochemical properties (valency and ion diameter) (Marschner, 1995). K^+ and Ca^+ are regarded to compete quite effectively with Mg^{2+} (Schimansky, 1981).

P of seedlings growing in sand/compost was low, which was only 33 % and 66 % of the content in seedlings growing in peat and sand, respectively. The low P content can be correlated with growth and leaf colour, resulting leaf area reduction (14 % lower than seedlings in peat and 24 % lower than those in sand) and a darker green colour of leaf, especially as compared with those growing in peat (figure 20 and table 6). On the other hand, the low P content did not affect root growth and shoot/root ratio of seedlings in sand/compost, especially as compared to seedlings growing in sand (figure 20). These results are accordance to Fredeen et al. (1989), who reported that, in contrast to shoot growth, root growth in soybean was less inhibited under phosphorus deficiency, leading to a typical decrease in shoot/root dry weight ratio.

3.4. Conclusions

"Pondoh hitam" is the most vigorous cultivar tested growing on different media such as sand, peat and sand/compost mixture. In general, the best growing media for salak seedlings growth is sand, supplied with a complete nutrition solution. Peat with a very low pH was not suitable as growing medium for salak seedlings due to the inhibitory growth effect on the seedlings. A high susceptibility of "pondoh super" seedlings to peat was found as compared to the other growing media tested. Net CO_2 assimilation rate (P_N) of salak of seedlings was relatively low as compared to other tropical tree species. Among the cultivars tested, "gading" exhibited the highest P_N. Foliar mineral contents (N, P, K, Ca and Mg) of salak seedlings were within the optimal range for growth in all media tested. However P content of plants growing in sand/compost was relatively low. Low P content of seedlings growing in sand/compost possibly induced leaf area reduction and darker green leaf colour.

4. THE EFFECT OF LIGHT AND WATER SUPPLY ON GROWTH, NET CO_2 ASSIMILATION RATE AND MINERAL CONTENT OF SALAK SEEDLINGS

4.1. Introduction

Environmental conditions, such as water availability and solar irradiation may become stress factors for plants if the dosage is too high or too low (Lüttge, 1997). However, the major factor in terms of plant stress is the complex interaction between plant and environment (Lüttge, 1997; Biswal and Biswal, 1999). Many environmental factors individually may not provoke any stress but in combination they can cause harmful situations for plants.

Water is among the most limiting factors for plant productivity and growth rate worldwide (Pugnaire et al., 1999), because of its essential role in plant metabolism, both at cellular and whole-plant level. Any decrease in water availability has an immediate effect on plant growth and physiological processes such as photosynthesis or mineral transport and accumulation (Hsiao et al., 1999). Water loss by transpiration causes transient water deficit, therefore the most plants suffer at least regular and daily water shortages. When drying, soil causes water absorption to lag behind transpiration and permanent water deficit may result in wilting and plant death due to dehydration (Pugnaire et al., 1999).

Photosynthesis is the fundamental physiological process for plant growth and production. Besides plant-related factors, environmental conditions influence net CO_2 assimilation rate (P_N), including irradiance, water availability, CO_2 concentration and temperature (Baswal and Baswal, 1999). Excessively high solar irradiance can damage the photosynthetic system, particularly in shade-adapted leaves or in leaves where the photosynthetic CO_2 assimilation has been inhibited by other stressors such as extreme temperature or water deficit (Jones, 1992). The damage can be a result of photooxidation where bleaching of chlorophyll occurs. Leaf injury not accompanied by bleaching is usually termed as photoinhibition (Jones, 1992).

Salak is an interesting palm tree growing at low light conditions in tropical rain forests of South-East Asia. In Indonesia, research on many aspects of salak fruit including pre- and postharvest have been carried out. Up to now, there is only limited knowledge about the physiological responses of salak to environmental factors. To promote production and fruit quality of salak, knowledge about ecophysiological aspects is very important. The purpose of this study was to investigate the effects of light and water supply on growth and physiological responses of salak seedlings.

4.2. Materials and Methods

4.2.1. Location and experimental design

The study was carried out from April 2003 to May 2003 for a period of 6 weeks in the greenhouse of the Department of Fruit Science in Berlin-Dahlem. The temperature of the greenhouse were adjusted to 20 °C/25 °C (day/night) and the relative humidity was between 40 % and 80 %.

Seven-months-old seedlings of salak cultivars "pondoh" from Indonesia were used for the study. At this age, plants are usually transferred from the nursery to the field in the growing region. The plant growing media was compost/sand (1/1, weight/weight). The compost media (pH 6) comprised of < 3 g/l NaCl, 450 mg/l K_2O and 320 g/l N and 320 mg/l P_2O_5 and was mixed with quartz sand (0.6 – 1.2 mm). Seedlings were planted in 14.5 cm x 11 cm pots. All plants tested were kept on greenhouse tables under shade nettings from germination until the beginning of the experiment. Shading was achieved using double shading net (Varia N2 9011, Wunder Lich, Germany), 2 m above the seedlings. Full sun light was reduced by 70 %.

An experimental block design was arranged for the study. A total of 48 experimental plants was separated into two blocks, namely shading (S) and non-shading (N). Three water supply treatments were assigned to each block. At the beginning of the study, the seedlings for N treatment were moved from shading to non-shading tables in the greenhouse. Two additional 400 W lamps (HQI-TS/D, OSRAM, Germany) were placed 2 m above the seedlings in the N treatment. Three watering treatments, applied to each block, were W1 (100 ml distilled water per plant every 2 days), W2 (100 ml per plant every 4 days) and W3 (100 ml per plant every 6 days). Therefore, the combinations of the treatments were SW1, SW2, SW3, NW1, NW2 and NW3.

Responses to be analysed were shoot and root dry weight, increment of shoot length and leaf area, net CO_2 assimilation rate (P_N) and plant mineral contents, i.e. nitrogen, phosphorus, potassium, calcium and magnesium. In addition to that, water content of the growing media was recorded.

4.2.2. Water content of the media

Water content of the media was measured with Time Domain Reflectometry (TDR), Soil Moisture Meter (70M/m/92, Easy Test Ltd., Poland) before the plants were watered according to the treatments. Digital readings ranged from 0 % (dry) to 100 % (saturated). During the study, 5 sets of TDR data were recorded (replicates of three media per treatment) from week 1 until week 5 of the experiment.

4.2.3. Growth

Growth parameter measurements were conducted similar to the procedure as described in Section 3.2.2.

4.2.4. Net CO_2 assimilation rate

Net CO_2 assimilation rate (P_N) of the leaves were measured at week 1, 2, 4 and 6 after the onset of treatments with a portable photosynthesis measuring system (CI-301PS, CID Inc., USA). The measurements were conducted starting from 10 am until 12 am, except at week 6 when measurements were taken from 11 am until 1.30 pm.

4.2.5. Mineral content

Mineral content of salak leaves were conducted similar to the procedure as described in Section 3.2.5.

4.2.6. Statistical analyses

All sample data were subjected to compare means and further analysed with least significant difference (LSD) test (Steel et al., 1996), with significant difference between means determined ($P \leq 0.05$) using the Statistic Program MINITAB 14 (Minitab Inc., State College, USA, 2001). The analyses comprised of solely the effect of light conditions or water supply treatments on growth, P_N and mineral content of salak seedlings. The effect of interactions between light conditions and water supply on growth, P_N and mineral content of salak seedlings was also analysed.

4.3. Results and Discussions

4.3.1. Water content of the media

Water content of the growing medium of salak seedlings grown under different light conditions are summarised in table 11. No differences in water content of salak seedlings media were found between shading and non-shading treatments up to week 3. From this time on, medium moisture of non-shading treatment (20.51 %) was always slightly lower than that of shading treatment (24.39 %).

Water content of the growing medium of salak seedlings grown under different water supply variants are presented in table 12. W1 represented a root zone moisture content of 30.83 % (29 - 33 %), W2 was 20.48 % (16 - 25 %) and W3 was 16.38 % (11 - 25 %).

Table 11. Water content of the growing medium of salak seedlings grown under different light conditions (shading (S) and non-shading (N))

Treatment	Week 1	Week 2	Week 3	Week 4	Week 5	Average
N (%)	17.47± 2.11 a	20.37± 2.20 a	21.23± 2.33 a	23.06± 1.23 a	20.42± 2.18 a	20.51
S (%)	21.01± 2.22 a	23.91± 1.21 a	23.66± 1.65 a	27.29± 1.14 b	26.09± 1.62 b	24.39

Values represent means ± SE. Different letters in the same column indicate significant differences by LSD test ($P \leq 0.05$)

Table 12. Water content of the growing medium of salak seedlings grown under different water supply (every 2 days (W1), every 4 days (W2) and every 6 days (W3))

Treatment	Week 1	Week 2	Week 3	Week 4	Week 5	Average
W1 (%)	31.15± 1.16 c	29.08± 0.67b	29.58± 0.94c	31.58± 1.03 c	32.78± 1.46 c	30.83
W2 (%)	15.75± 0.77 b	19.54± 1.55a	25.19± 1.15b	20.80± 0.97 a	21.12± 0,92 b	20.48
W3 (%)	10.81± 0.49 a	16.71± 1.55a	13.93± 1.08a	24.56± 0.65 b	15.87± 1.66 a	16.38

Values represent means ± SE. Different letters in the same column indicate significant differences by LSD test ($P \leq 0.05$)

4.3.2. Growth

The effect of light on growth parameters of salak seedlings are summarised in figure 24 and table 13. Root and shoot dry weight as well as the increment of leaf area of seedlings growing in non-shading conditions were higher as compared with those in the shading variant (46 %, 31 % and 14 % respectively). Similar results were reported by Stoneman and Dell (1993) and Stoneman et al. (1995) on *Eucalyptus marginata* seedlings, i.e. growth of all plant parts increased in response to higher light level. In grape (*Vitis vinifera*), leaf, shoot, trunk, root and total plant dry weights were reduced when growing under lower light level (Van den Heuvel et al., 2004); root growth decreased by 84 % in vines that had been shaded for a period of 16 weeks (Van den Heuvel, 2002). Root dry weight of vines growing under 80 % shading were reduced by 51 % in the following year (McArtney and Ferree, 1999). The increment of shoot length of seedlings in the non-shading variant was smaller as compared with those under shading conditions (29 %). This can be explained as etiolation effect indicating that under the shading net, light intensity was too low for proper growth of salak seedling.

The effect of water supply on growth of salak seedlings is presented in table 14. No differences of growth parameters were found among water supply treatments. However, there was a strong effect of water supply on leaf area, which is the most important parameter. The reduction by 26 % in the increment of leaf area of both W2 (watered every 4 days) and W3 (watered every 6 days) treatments as compared with W1 (watered every 2 days) was found. The reduction of leaf area appears to be largely affected by soil water status and by root hydration (Termaat et al., 1985). This could be correlated with the result of water content of the media mentioned in Section 4.3.1., that more frequent watering resulted in higher water content of the media. On the other hand, Levitt (1980) stated that decreased cell turgor (cell enlargement) is the most sensitive plant response to water stress, since cell growth is quantitatively related to cell turgor, and cell turgor decreases with any dehydration-induced decrease in cell water potential. Factors which postpone dehydration by reducing water loss, such as a decrease in leaf area, are known to decrease productivity (Turner, 1979). Moreover, it was found, that salak plants are drought susceptible, since the plants began to wilt during a 2 to 3 weeks drought period (Lestari and Ebert, 2002)

The growth parameters affected by light and water supply treatments are summarised in table 15. Seedlings growing in full light with highest water supply had the highest root and shoot weight as well as leaf area increment. This result indicated non-shading and high water supply were the most suitable conditions for growth of salak seedlings. On the other hand, seedlings of the shading variant and high water supply developed the lowest root and shoot dry weight. Shading in the study obviously reduced too much light, which was required by seedlings for growing more vigorously. On the other hand, more frequent water supply and shading caused excessive high water content of the growing media which possibly lead to oxygen depletion. These conditions are regarded to cause the inhibition of leaf extension and cessation of root growth (Marschner, 1995).

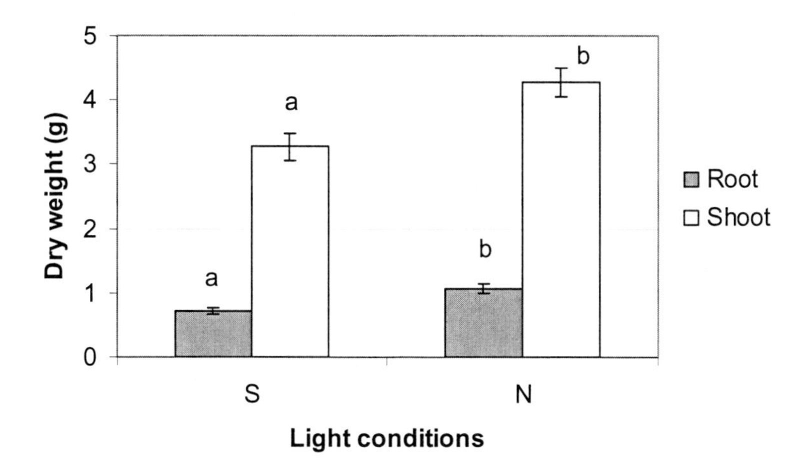

Figure 24. Root and shoot dry weight of salak seedlings under different light conditions (shading (S) and non-shading (N))

Different letters in the same column indicate significant differences by LSD test ($P \leq 0.05$)

Table 13. Leaf area and shoot length increments of salak seedlings under different light conditions (shading (S) and non-shading (N))

Growth parameter	Light condition	
	S	N
Leaf area (cm^2)	150.69 ± 19.40 a	171.99 ± 17.6 a
Shoot length (cm)	10.14 ± 1.56 a	7.15 ± 1.25 a

Values represent means ± SE. No significant differences by least significant difference (LSD) test ($P \leq 0.05$) were found.

Table 14. Growth of salak seedlings under different water supply (every 2 days (W1), every 4 days (W2) and every 6 days (W3))

Growth parameter	Water supply		
	W1	W2	W3
Root (g DW)	0.89 ± 0.11 a	0.91 ± 0.08 a	0.89 ± 0.10 a
Shoot (g DW)	3.85 ± 0.34 a	3.76 ± 0.35 a	3.73 ± 0.24 a
Leaf area (cm^2)	195.26 ± 24.60 a	143.79 ± 23.90 a	144.96 ± 17.00 a
Shoot length (cm)	6.31 ± 2.09 a	11.00 ± 1.69 a	8.63 ± 1.30 a

Values represent means ± SE. No significant differences by least significant difference (LSD) test ($P \leq 0.05$) were found.

Table 15. Interactive effects of different light conditions (shading (S) and non-shading (N)) and water supply (every 2 days (W1), every 4 days (W2) and every 6 days (W3)) on growth of salak seedlings

Treatments	Growth parameter			
	Root (g DW)	Shoot (g DW)	Leaf area (cm^2)	Shoot length (cm)
SW1	0.65 ± 0.10 a	3.00 ± 0.29 a	165.05 ± 36.40 a	9.77 ± 3.68 a
SW2	0.81 ± 0.11 ab	3.34 ± 0.53 ab	141.90 ± 48.20 a	10.73 ± 2.51 a
SW3	0.72 ± 0.07 ab	3.48 ± 0.22 ab	145.11 ± 11.70 a	9.92 ± 2.23 a
NW1	1.12 ± 0.15 b	4.70 ± 0.35 b	225.48 ± 31.30 a	2.85 ± 0.93 a
NW2	1.01 ± 0.11 ab	4.17 ± 0.43 ab	145.69 ± 13.80 a	11.27 ± 2.49 a
NW3	1.05 ± 0.16 ab	3.98 ± 0.42 ab	144.80 ± 33.70 a	7.33 ± 1.34 a

Values represent means ± SE. Different letters in the same column indicate significant differences by LSD test ($P \leq 0.05$)

4.3.3. Net CO_2 assimilation rate

Net CO_2 assimilation rate (P_N) of salak seedlings under different light conditions is presented in figure 25. Under non-shading (N) condition during the measurements photosynthesis active radiation (PAR) was about 400 to 700 µmol m^{-2} s^{-1}, whereas shading (S) reduced PAR by 70 %. Until week 4, P_N of seedlings in non-shading (varying in between 3.4 and 3.6 µmol CO_2 m^{-2} s^{-1}) was higher as compared with in shading condition (varying from 2.2 to 3.0 µmol CO_2 m^{-2} s^{-1}). This result gave evidence that shading limited P_N in salak seedlings. Higher light intensity led to higher P_N in salak seedlings of non-shading condition as compared with that of seedlings under a shading net, except the midday P_N measurement at week 6. A great number of publications give evidence, that photosynthetic CO_2 – assimilation is limited by light intensity (Wong et al., 1978; von Caemmerer and Farquhar, 1981; Stoneman et al., 1995). Higher P_N resulted in more chemical energy in the form of carbohydrates and other organic compounds, which are ultimately used for plant growth. Similar results have been reported for banana plants, where P_N was lower under shading conditions as compared to plants in full sun light (Thomas and Turner, 2001). This was due to the limitation of P_N by low photosynthetic photon flux density. The mesophyll resistance, which is usually regarded as the sum of biophysical and biochemical resistances to CO_2 movement between the mesophyll cell wall and the site of carboxylation in the chloroplast, was reported to be higher in plants grown at low light intensities (Holmgren, 1968, Crookston et al., 1975).

There was no net CO_2 gain in non-shaded plants at week 6, measured over midday hours (figure 25). The phenomenon of low P_N at intensive solar irradiation and high water vapour saturation deficit is well known for many plants species ("midday depression"). In our experiment, this effect was induced by high light radiation (up to PAR of 1636 µmol m^{-2} s^{-1}) during the P_N measurement. The "midday depression" in plants subjected to water stress is associated with photoinhibition. Both light and water stress evidently initiate an activation of the primary photochemical pathway associated with photosystem II (Björkman and Powles, 1984). This could be proved by the occurrence of bleaching in some salak leaves subjected to full sun light. It is well established that photobleaching is a secondary phenomenon following photoinhibition

(Powles, 1984). Similar photoinhibitory effects or midday reduction of P_N at high photon flux density (more than 1000 µmol m^{-2} s^{-1}) have been reported for water-stressed carambola (Ismail et al., 1994) and citrus plants (Brakke and Allen Jr., 1995). This was typically attributed to large leaf-to-air vapour pressure difference or high atmospheric vapour pressure deficits. Therefore, in the case of light-susceptible salak, shade is an important factor for the survival of plants especially with respect to water stress. As known from a study, salak palms are very susceptible to high light intensities (Lestari and Ebert, 2003).

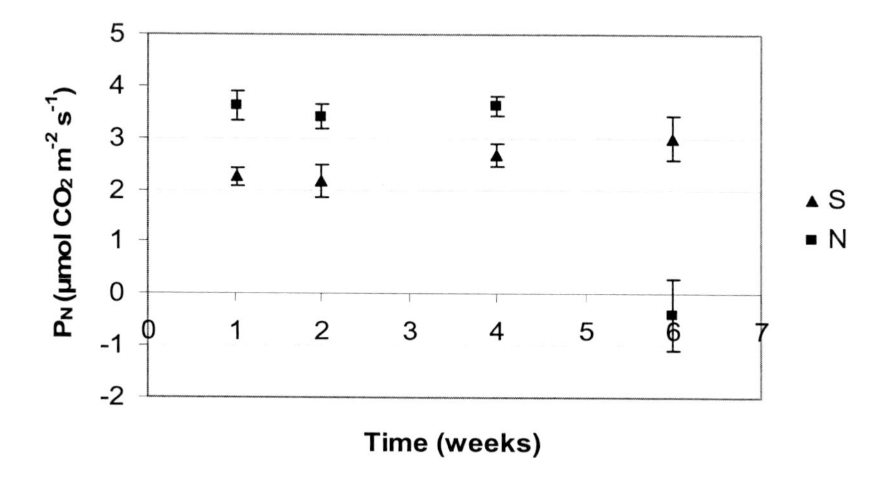

Figure 25. P_N of salak seedlings under different light conditions (shading (S) and non-shading (N))

P_N of salak seedlings at different water supply is presented in figure 26. Until week 4 of the treatment, there was no clear influence of water supply on P_N (ranging from 2.5 to 3.4 µmol CO_2 m^{-2} s^{-1}). However, P_N decreased at week 6, in all water supply treatments. Moreover, in some P_N measurements at week 6, there was no net CO_2 gain in W3 plants, but only leaf respiration. Some studies have suggested that the change in stomatal conductance was the main cause for the decrease in P_N in water-deficient plants (Boyer, 1976; Cornic, 2000). In *Nerium oleander* plants, subjected to water stress, full sun light resulted in an inhibition of non-stomatal components of photosynthesis. This was manifested as a reduced photon yield and light-saturated capacity of photosynthetic CO_2 uptake, measured after restoration of high leaf water potential and partial reopening of the stomata (Björkman and Powles, 1984). Either stomatal closure or non-stomatal limitation might have induced midday depression of P_N as observed in non-shaded salak plants. In case of stomatal closure, CO_2 supply from ambient air becomes the limiting step in the CO_2 assimilation process (Pugnaire et al., 1999).

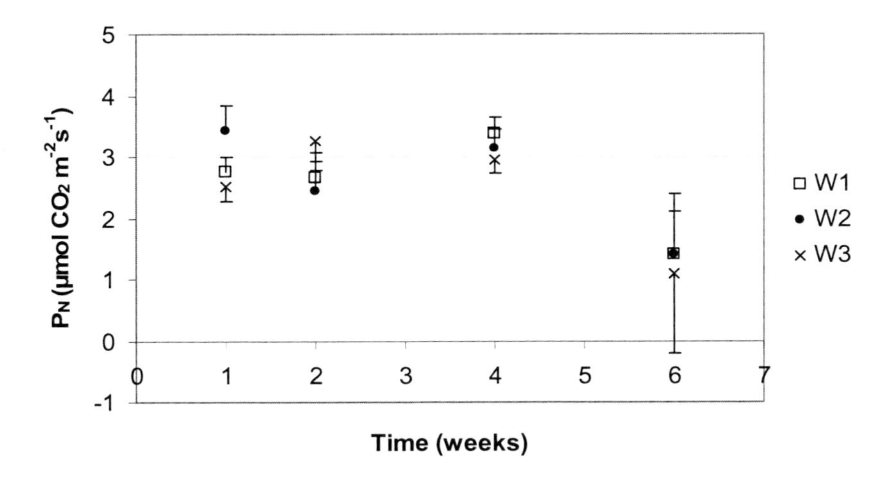

Figure 26. P_N of salak seedlings under different water supply (every 2 days (W1), every 4 days (W2) and every 6 days (W3))

Table 16. Interactive effects of different light conditions (shading (S) and non-shading (N)) and water supply (every 2 days (W1), every 4 days (W2) and every 6 days (W3)) on net CO_2 assimilation rate of salak seedlings

Treatments	P_N (μmol CO_2 m^{-2} s^{-1})			
	Week 1	Week 2	Week 4	Week 6
SW1	2.76 ± 0.33 a	2.43 ± 0.70 ab	3.05 ± 0.45 ab	3.05 ± 0.45
SW2	2.18 ± 0.25 a	1.53 ± 0.39 a	2.43 ± 0.40 a	2.57 ± 0.84
SW3	1.88 ± 0.25 a	2.56 ± 0.46 ab	2.62 ± 0.29 ab	3.32 ± 0.91
NW1	2.74 ± 0.38 a	2.88 ± 0.47 ab	3.72 ± 0.21 ab	nd
NW2	4.82 ± 0.53 b	3.37 ± 0.37 b	3.87 ± 0.40 b	nd
NW3	3.31 ± 0.29 a	3.95 ± 0.34 b	3.30 ± 0.36 ab	nd

nd = not detected

Values represent means ± SE. Different letters in the same column indicate significant differences by LSD test ($P \leq 0.05$)

P_N of salak seedlings affected by interaction of light and water supply are presented in table 16. P_N of plants in all water supply treatments with non-shading condition tended to be higher than that of the interactions with shaded plants. P_N of NW2 plants at week 1 and 4 was found to be the highest. The midday measurement of week 6 indicated that plants under full sun light failed to assimilate CO_2. The results indicate that light had a stronger effect on P_N in salak seedlings than water supply treatment. Some studies also reported that "midday–depression" of P_N under high irradiance can be observed either in stressed or non-stressed young carambola plants (Ismail et al., 1994), in well-irrigated *Protea acaulos* (Herppich et al., 1994) and in young, well-watered *Welwitschia mirabilis* (Herppich et al., 1996).

P_N of salak is generally low, even at high PAR. This phenomenon is a characteristic of shade plants, which have very low light compensation points i.e. 0.5 – 2 μmol m^{-2} s^{-1} in the extreme shade species *Alocasia macrorrhiza* (Jones, 1992). At high PAR, P_N of some other tropical tree species, such as peach palm (*Bactris gasipaes* Kunth) is about 10 μmol CO_2 m^{-2} s^{-1} (Oliveira et al., 2002), oil palm 25 μmol CO_2 m^{-2} s^{-1} (Dufrene and Saugier, 1993), mango 8 μmol CO_2 m^{-2} s^{-1} (Fukamachi et al., 1998), banana is 19 μmol CO_2 m^{-2} s^{-1} (Thomas and Turner, 2001) and citrus is 11 CO_2 m^{-2} s^{-1} (Brakke and Allen Jr., 1995).

4.3.4. Mineral content

N content of leaves was strongly influenced by light conditions. There was no influence on the uptake of other ions (table 17). Estimates of the quantity of nutrients supplied to plants by mass flow are based on not only the nutrient concentration in the soil solution but also the transpiration (Marschner, 1995). Movement of nutrients (such as NO_3) to the root surface via mass flow and subsequently from the root into leaves are predominantly influenced by the transpiration (Shaner and Boyer, 1976). Higher transpiration rates under full-light condition will result in higher nutrient uptake. These could possibly explain the increased nitrogen concentration in the leaves of non-shaded plants as compared to that of shaded plants. This result might also relate with reduced growth and lower P_N of salak seedlings in shading as compared with in non-shading conditions (figure 24, table 13 and figure 25). A strong relationship exists between total plant nitrogen concentration and the fraction of plant dry matter that is produced in leaf tissue (Agren and Ingestand, 1987). Furthermore, Agren (1985) and Hirose (1988) modelled the plant relative growth rate as a linear function over a wide range of whole-plant internal nitrogen concentrations. There is also a tight relation between leaf nitrogen and maximum photosynthetic activity (Le Bot et al., 1998). Mathematical functions have been applied for modelling the effect of nitrogen on P_N (DeJong and Doyle, 1985; Hirose and Werger, 1987; Sage and Pearcy, 1987; Hirose at al., 1988; Lim et al., 1990). Since the growth of plants is a function of their photosynthesic capacity and allocation to leaf tissue, interrelationships between nitrogen productivity, leaf growth, net assimilation rate and relative growth rate can be modelled (Agren and Ingestand, 1987; Hirose, 1988)

On the other hand, there were no differences in mineral contents of salak seedlings in all water supply treatments (table 18). This indicated that water supply did not limit the nutrient uptake of salak seedlings.

The contents of N, P, K, Ca and Mg in salak seedlings affected by interaction of light and water supply are summarised in table 19. Nitrogen content of SW1 plants was lower than that of NW2 and NW3 plants, but not different to that of the other interactions tested. On the other hand, no differences of nitrogen content in salak leaf were found between similar water supply treatment in non-shading and shading condition. These results indicated that light had a stronger effect on nitrogen content of salak seedlings as compared to water supply. Interactions of light and water supply did not affect mineral contents of salak seedlings in this experiment.

According to Marschner (1995), the optimal ranges of mineral content in plant shoot for growth (in mg/g) are as follows:

N = 20 – 50; P = 3 -5; K = <20 – 50, Mg = 1.5 – 3.5 and Ca = 1 - > 50

Optimal leaf element concentration for oil palm (*Elaeis guinensis*) according to Fairhorst and Härdter (2003) as follows (in mg/g):

N = 14   P = 1.5 – 1.9   K = 10 – 13
Mg = 1.6 – 2.2   Ca = 1.4 – 2.5

All experimental plants showed nutrient contents being in the optimal range (Marschner, 1995), with the exception of Mg. This might be due to a limited availability of this element in the media, since there was no supply of nutrients to the seedlings during the experiment. Only N was found to be significantly lower in shaded plants. The results clearly indicate that under the experimental condition as explained above, nutrient uptake of salak seedlings was neither affected by shading nor by different lack of water supply.

Table 17. Mineral content of salak seedlings under different light conditions (shading (S) and non-shading (N))

Light condition	N	P	K	Ca	Mg
			(mg/g DM)		
S	25.39 ± 0.3 a	2.78 ± 0.1 a	13.95 ± 0.5 a	4.28 ± 0.2 a	1.01 ± 0.03 a
N	27.27 ± 0.4 b	2.60 ± 0.1 a	12.51 ± 0.7 a	5.49 ± 0.6 a	1.10 ± 0.05 a

Values represent means ± SE. No significant differences by LSD test ($P \leq 0.05$) were found

Table 18. Mineral content of salak seedlings under different water supply (every 2 days (W1), every 4 days (W2) and every 6 days (W3))

Water supply	N	P	K	Ca	Mg
			(mg/g DM)		
W1	25.58 ± 0.6 a	2.74 ± 0.06 a	12.88 ± 1.07 a	3.85 ± 0.20 a	1.07 ± 0.05 a
W2	26.64 ± 0.7 a	2.83 ± 0.15 a	12.80 ± 0.85 a	5.30 ± 0.56 a	1.04 ± 0.06 a
W3	26.76 ± 0.6 a	2.49 ± 0.09 a	14.01 ± 0.22 a	5.50 ± 0.68 a	1.06 ± 0.04 a

Values represent means ± SE. No significant differences by LSD test ($P \leq 0.05$) were found

Table 19. Interactive effects of different light conditions (shading (S) and non-shading (N)) and water supply (every 2 days (W1), every 4 days (W2) and every 6 days (W3)) on mineral content of leaves of salak seedlings

Interaction	N	P	K	Ca	Mg
	(mg/g DM)				
SW1	24.98 ± 1.0 a	2.78 ± 0.05 a	14.74 ± 1.0 a	3.70 ± 0.35 a	1.04 ± 0.07 a
SW2	25.48 ± 0.1 ab	3.05 ± 0.09 a	13.21 ± 1.0 a	4.60 ± 0.22 a	1.00 ± 0.02 a
SW3	25.72 ± 0.2 ab	2.50 ± 0.03 a	13.90 ± 0.4 a	4.54 ± 0.47 a	1.00 ± 0.06 a
NW1	26.19 ± 0.6 ab	2.70 ± 0.13 a	11.03 ± 1.4 a	4.00 ± 0.23 a	1.10 ± 0.08 a
NW2	27.81 ± 0.4 b	2.61 ± 0.16 a	12.39 ± 1.5 a	6.00 ± 1.04 a	1.09 ± 0.11 a
NW3	27.81 ± 0.2 b	2.48 ± 0.21 a	14.12 ± 0.2 a	6.47 ± 1.14 a	1.12 ± 0.05 a

Values represent means ± SE. No significant differences by LSD test ($P \leq 0.05$) were found

4.4. Conclusions

From the study it is concluded that shading, which reduced sunlight by 70 %, was not beneficial for 7-months-old salak seedlings, due to lower growth rate, P_N and N content as compared with plants in non-shading conditions. Six weeks after the treatments, photoinhibitory response of non-shaded plants was detected, which caused bleaching of salak leaves. Shading is to a certain extent required for raising salak seedlings. However, light intensities less than 250 µmol m^{-2} s^{-1} (PAR) may lead to retarded growth, decreased P_N and inhibited nitrogen uptake. Light intensities above 800 µmol m^{-2} s^{-1} (PAR) possibly cause leaf bleaching and should be avoided. Salak plants did not tolerate drought conditions. However, different water supply did not affect P_N and mineral uptake (N, P, K, Ca, Mg) of salak plants. Light had a stronger effect on P_N and nutrient uptake of salak seedlings as compared to water supply.

5. FRUIT QUALITY OF DIFFERENT SALAK CULTIVARS (*Salacca zalacca* (Gaertn.) Voss): EXTERNAL PROPERTIES, NUTRITIONAL VALUES AND SENSORY ATTRIBUTES

5.1. Introduction

Quality is defined as the sum of characteristics, properties and attributes of a product or commodity which is aimed to fulfil the established or presumed customer requirements (DIN ISO 8402, 1989). Quality attributes of fresh fruits include appearance, texture, flavour and nutritive value (Kader and Barrett, 1996). Appearance includes parameters such as size, shape, colour and absence of defects, whereas textural properties are described by e.g. firmness, crispness and juiciness. Flavour components incorporate sweetness, acidity, astringency, bitterness, aroma and off-flavours. Nutritional quality is determined by the content of carbohydrates, protein, fat, minerals, dietary fibres, vitamins and other components with beneficial effects on consumer's health.

Quality is not only a single attribute, but a combination of characteristic. Shewfelt (1999) suggests that the combination of characteristics of the product itself should be termed quality and that the consumer´s perception and response to those characteristics should be referred to as acceptability. There is no single, universal consumer type, as consumers have quite various requirements (Huyskens-Keil and Schreiner, 2003)

The market and demand for new exotic subtropical and tropical fruits, revealing a high nutritional and sensory value, has significantly increased during the past few years in the developed countries (ZMP-Markbilanz Obst, 2000). Among these crops, the salak fruit possesses a prospective export potential. To promote the entry of salak into the worldwide fruit market, the key quality aspects of the fruit need to be clarified.

The purpose of this study was to determine, to compare and evaluate the fruit quality of different salak cultivars "pondoh super", "pondoh hitam", "pondoh manggala" and "gading" with special emphasis on external properties, nutritional value and sensory attributes.

5.2. Materials and Methods

5.2.1. Plant material

Four salak cultivars from Sleman, Yogyakarta, Indonesia were used for this investigation: "gading", "pondoh super", "pondoh hitam" and "pondoh manggala" (figure 27). Fruits were harvested in October 2002, i.e. 4.5 months to 5.5 months after pollination. Immediately after air transport from Indonesia to Germany, fruit quality properties were assessed in Berlin and Quedlinburg. Part of the fruit material was deep-frozen. For the determination of nutritional properties, freeze-dried fruits of salak were used. The fruit samples for GC-MS analysis of aroma compounds had been stored under deep-freeze conditions for 4 months. For gas chromatography-olfactometry (GCO) determination, fresh fruit samples, harvested in June 2003 at stage 5.5 to 6 months after pollination, of cultivars "gading", "pondoh super" and "pondoh hitam" were used. The fruit developmental stages analysed in this study represent those, which are usually presented on the local markets in Indonesia.

5.2.2. External properties

Fresh fruits (n =15 of each cultivar) were weighed and the edible portion was determined. The size of the fruits (length and diameter) was measured using callipers. Dry weight was determined after drying the sample at 103 °C for 24 hours until constant weight (Maier, 1990). Measurement of fruit peel and pulp colour was conducted using a Minolta Colorimeter (CR-321, Minolta, Japan) as described in section 3.2.4. Peel colour was recorded on the equator region of each fruit with eight measurements. Pulp colour was computed on the tip and base regions of each fruit with four measurements. Pulp firmness was determined non-destructively using a Shore A instrument (HHP-2001, Bareiss, Germany) equipped with a 2.5 mm diameter probe and a standardised metal disc. Measurements were carried out on 3 equidistant records of the equatorial fruit region and were expressed as Shore A units.

Figure 27. Salak fruits of different cultivars ("gading" (upper left), "pondoh super" (upper right), "pondoh hitam" (lower left) and "pondoh manggala" (lower right))

5.2.3. Nutritional valuable properties

5.2.3.1. Soluble solid content and titratable acidity

Soluble solid content (SSC) was measured after extracting the juice of 4 fruits per cultivar, with two replicates. SSC was measured using a digital refractometer (model PR 101, Kuebler, Germany). A calibration with distilled water was conducted before each measurement. Refractometric reading was expressed as index of refraction (°Brix) at 20 °C (AOAC, 1990). Measurement of the titratable acidity was carried out by titration of juice samples with 0.1 N NaOH up to pH 8.1 (AOAC, 1990) and expressed as malic acid, the dominant organic acid in salak fruit (Hartanto, 1998). The sugar/acid ratio was calculated as SSC/titratable acidity.

5.2.3.2. Minerals

500 mg freeze-dried fruit sample triplicates of each cultivar were used for the determination of K, Ca, Mg and P. The analysis was conducted as described in section 3.2.5. The data of the minerals were expressed as mg/g DM.

5.2.3.3. Fat

Fat of salak fruits was analysed according to the method described by Stoldt (1952). Replicates of 1g freeze-dried samples were hydrolysed by adding 100 ml 4 N HCl and then boiled for 20 min. The solution was filtered and the residue was washed with boiling water to remove the acids (pH 6-7). Thereafter, the samples were dried at 98 °C for 90 min and were transfered into the extraction cups. Samples were extracted for 1 h with 50 ml petroleum ether solvent (boiling temperature 40 to 60 °C) (Soxtec System HAT, Tecator, Sweden). After the extraction process, the solvent was evaporated. Subsequently, the residues were dried at 105 °C for 1 h and were weighed thereafter. The data were calculated by the difference of the weight before and after extraction and were expressed in % DM.

5.2.3.4. Mono- and disaccharides

Mono- and disaccharides (glucose, fructose and sucrose) were determined by High Performance Liquid Chromatography (HPLC) (Ulrichs, 1999). Analyses were performed in triplicates per cultivar and the data were expressed in mg/g DM. Freeze-dried samples (100 mg) were added with 80 % ethanol, and placed in a stirring water bath at 70 °C for 20 minutes. Thereafter, the samples were centrifuged at 3000 rpm for 15 min. The supernatant was collected in a volumetric flask, whereas pellets were stirred with 80 % ethanol. Extraction with ethanol was conducted three times, thereafter HPLC water was added to the samples. The supernatants were evaporated in a rotavapor

(RE 120, Buechi, Switzerland). A defined volume of HPLC water and saturated lead acetate solution were added to the dried samples, which were centrifuged thereafter at 3000 rpm for 20 min. The supernatant was transferred to an Eppendorf-funnel with ion exchanger V (Merck) and shaken for 30 min. Finally, the samples were purified in extract clean columns (C18, Alltec, Germany) and kept at –30 °C until further analysis (HPLC).

A HPLC instrument (Model 25, Fa. Bischoff, Germany), equipped with an autosampler (Alcott 708) and a RI-detector (8110, Fa. Bischoff, Germany) was used. A Waters-Spherishorb Amino (250 mm x 3.0 mm) column was used with a 3 μm-packing material. The mobile phase was acetonitrile/water (85:15). HPLC was operated at an ambient temperature with a flow rate of 1 ml/min. 10 μl sample was injected into the autosampler. Quantification was accomplished by determination of the area under the chromatographic peak and calculation of the level of each component on the basis of standard curves generated with pure compounds. Standard sugar solution (Fa. Seroa and Boehringer, Germany) contained 10 μg/10 μl of both for fructose and glucose, and 25 μg/10 μl of sucrose.

5.2.3.5. Cellulose, hemicellulose and lignin

Cellulose, hemicellulose and lignin were analysed according to the method of Goering and Soest (1972) and AOAC (1984). 1 g freeze-dried sample material was extracted with 100 ml Neutral Detergent Fibre (NDF) reagent or Acid Detergent Fibre (ADF) reagent using the hot extractor unit (Fibertec System M 1020, Tecator, Sweden) to get content of NDF or ADF respectively. NDF reagent comprised of 18.61 mol/l EDTA, 6.81 mol/l natriumtetraborat, 30 mol/l laurylsulphate, 10 ml/l ethyleneglycolmonoethylether and 4.56 mol/l dinatrium-hydrogenphosphate. ADF reagent comprised of 20 g cetyltrimetylammoniumbromide, which was adjusted to 11 by adding 1mol/l H_2SO_4. After extraction, the solution was vacuum filtered, washed with boiled water until acid removal and again washed with 90 % acetone. NDF and ADF residues were dried at 105 °C for 24 h, weighed, ash-dried at 500°C for 24 h and weighed again to calculate NDF and ADF contents.

ADF residue served as material for further lignin determination. The residues were again submerged with 72 % H_2SO_4 for 3 h at room temperature (20 to 23 °C), washed with hot water until acid removal and washed again with 90 % acetone. Subsequently, the material was dried at 105 °C for 24 h, continued by ash-drying at 500 °C for 24 h.

NDF, ADF and lignin were calculated from the ratio before and after the samples have been ash-dried and the value was expressed in % DM. The content of hemicellulose was calculated by the difference between NDF and ADF, whereas the content of cellulose was calculated by the difference between ADF and lignin content.

5.2.3.6. Pectic substances

Cell walls were prepared following the method described by McComb and McCready (1952), Blumenkrantz and Asboe-Hansen (1973) and Huyskens (1991). Triplicates of 1.5 g freeze-dried samples were mixed with 100 ml 99.9 % acetone and boiled for 30 min. Thereafter, the suspension was vacuum filtered. The residue on the filter paper (Schleicher and Schuell No 589/3) was resuspended in 99.9 % acetone, 70 % ethanol and again in 99.9 % acetone to remove mono- and disaccharides. The final white residue on the filter paper was dried overnight at 70 °C. The alcohol insoluble solids (AIS) fraction was weighed and stored in a vacuum desiccator.

AIS were fractionated into water soluble pectin (WSP), ethylene diamin tetra acetic acid-soluble pectin (EDTA-SP) and insoluble pectin (ISP) according to the method of Blumenkrantz and Asboe-Hansen (1973). For the determination of WSP, AIS was mixed with 20 ml distilled water using a magnetic stirrer (IKA-Labortechnik, Germany) at room temperature for 1 h. The solution was brought to pH 4.5 by adding 1:1 diluted acetic acid. Then 0.1 ml pectinase (ca. 20 µg) (Pectinex Ultra SP-L, Novo Nordish Ferment Ltd, Ditlingen, Switzerland) was added to the solution, which was stirred for 1 h. The solution was placed in 100 ml centrifugal tubes and centrifuged (Biofuge 22R, Heraeus Sepatech, Germany) at 4 °C at 11.000 rpm for 10 min. The supernatant and the pellets were filtered through Miracloth in a 50 ml volumetric flask. The flask was filled with 0.5 % EDTA-solution (pH 4.5) up to 50 ml. The supernatant was refiltered through a filter paper (Schleicher and Schuell No 589/3). The remaining pellet was kept in deep freezes for further extraction of the EDTA-SP fraction. For this extraction, a similar procedure as for water soluble pectin fraction was applied. Instead of mixing the AIS samples with 20 ml distilled water, the pellet of the WSP fraction was mixed with 20 ml 0.5 % EDTA (pH 6.0). The remaining pellet at the end of procedure was kept for further extraction of ISP. For this extraction, the pellet of the EDTA-SP fraction was mixed with 20 ml 0.5 % EDTA (pH 11.5) and stirred for 30 min. After 10 min, the pH was adjusted to 11.5.

The pectin fractions WSP, EDTA-SP and ISP were determined as described by McComb and McCready (1952). One aliquot of 0.7 ml, 0.5 ml and 0.2 ml filtrate of each fraction sample (WSP, EDTA-SP and ISP, respectively) was filled up to 1 ml by adding 0.5 % EDTA-solution (pH 6). Then 6 ml of ice-cold concentrated sulphuric acid were added to the samples to solubilise the polysaccharides. The samples were boiled in a water bath at 100 °C for 10 min to hydrolyse the polysaccharides to monomeric sugars and then cooled-down to room temperature. MHDP-solution 0.1 ml 0.15 % was added to the sample. The pectin content was determined spectrophotometrically at a wavelength of 520 nm (UV/VIS-8730, Phillips, UK). Standard solutions of D-galacturonic acid were used for the calibration. Data were expressed as mg galacturonic acid/g DM.

5.2.4. Sensory attributes

For the evaluation of sensory attributes, a descriptive consumer (untrained) panel test (n = 5) was carried out and quality attributes of salak (general appearance, external and internal colour, easy-to peel, aroma, texture (hard-soft), textural acceptance, sweetness, acidity, taste acceptance, taste intensity and undesired aftertaste) were judged using a scale from 0 to 5 (0 = no value, 1 = very weak, 2 = weak, 3 = medium intensive, 4 = intensive and 5 = very intensive) (Jellinek, 1985).

5.2.5. Aroma compounds

Aroma compounds of salak fruits were analysed according to the method as described by Wijaya et al. (2004).

5.2.5.1. Isolation of volatiles by liquid-liquid extraction

The pulp (200 g) of salak fruit was homogenised for 1 min in 300 ml NaCl solution (18.6 % w/v). The homogenate was centrifuged at 4 °C for 30 min at 3000 rpm. To obtain a clear liquid, the supernatant was filtered through filter paper. A portion (250 ml) of the filtrate was subjected to a fluid-fluid extractor apparatus (Rapp et al., 1976) after addition of an internal standard (10 µl of 2,6-dimethyl-5-heptenol per 100 ml sample solution). Volatile compounds were extracted with 30 ml 1,1,1-trichlorflourmethane (Freon F11) from the aqueous solution for 20 hours at room temperature. The freon fraction was concentrated with a Vigruex column directly before analyses.

5.2.5.2. Isolation of volatiles by stir bar sorptive extraction (SBSE)

An aliquot of 10 ml supernatant prepared as described above was pipetted into a 20 ml glass headspace vial and saturated by adding 3 g of solid NaCl. A stir bar (0.5 mm film thickness, 10 mm length, Gerstel GmbH, Mühlheim/Ruhr, Germany) coated with poly(dimethylsiloxane) (PDMS) was used. The stir bar was placed in the 20 ml vial, which was sealed with a crimp cap and stirred at 300 rpm at room temperature for 45 min. After removal from the sample, the stir bar was rinsed with distilled water, dried with a tissue and transferred in a glass thermal desorption tube for GC-MS analysis.

5.2.5.3. Gas chromatography - mass spectrometry (GC-MS)

For liquid samples a Hewlett-Packard GC-MS system (GC 5890 plus and MSD 5972) equipped with split-splitless injector at 250 °C was used. The GC-MS was equipped with a Gerstel Thermal Desorption Unit (TDU) and Cooled Injection System 4 (CIS4).

The MS detector temperature was 280 °C. A polar column (HP INNOWax, 0.25 mm ID x 30 m length x 0.5 µm film thickness) was used with the following temperature program: 45 °C held for 5 min, then raised to 200 °C at a rate of 2 °C/min and held for 30 min. The flow rate of the carrier gas (He) was 1 ml/ min. A volume of 1 µl of each sample was injected with a split ratio of 1:3, 1:10 1:25, respectively. For compound identification, the HP Chem Station software with Wiley/National Bureau of Standards (NBS) library and the National Institute of Standards and Technology (NIST) library were used.

5.2.5.4. Gas chromatography olfactometry (GCO)

A Hewlett Packard GC 6890 series II gas chromatograph equipped with a polar fused silica column (HP INNOWax, 0.25 mm ID x 15 m length x 1.0 µm film thickness) was used with split injection. Both injector and flame ionisation detector (FID) temperatures were 250 °C. The oven program was the following for all runs: 45 °C for 5 min, then rose to 210 °C at a rate of 10 °C/min, and finally held for 5 min. Each sniffing run continued for 25 min. Hydrogen was used as carrier gas at a flow rate of 1 ml/min. A volume of 1 µl of salak fruit extract was injected at a split flow of 5 ml/min. The column outlet was connected either with the FID via transfer line (monitor run) or with the sniffing port equipped with a heated transfer line of identical length and a glass funnel. The complete sniffing session for one cultivar consists of two FID analyses, one with a boiling point sample for calculation of retention indices and one monitor run with the salak extract for retention time adjustment to the GC-MS run of the same extract. After switching the column outlet to the sniffing port, six separate runs were carried out by six trained panellists.

To determine potentially odour-active compounds of the salak fruit extracts, the nasal impact frequency (NIF) method (Pollien et al., 1997) was performed with modifications (sniff.exe software). Panellists recorded the perception of an odour by pressing a button as long as the smell could be received. Moreover, each panellist was encouraged to describe the odour quality of each perception. The signals were registered by the HP Chemstation via an A/D converter as square signals. Following the six single sniffing runs were accumulated to one coincident response chromatogram called NIF profile (Pollien et al., 1997) using a specially designed macro (sniff.exe software). The resulting NIF profile was handled by the HP Chemstation to integrate peaks. Odours detected only by 1 or 2 judges were considered as olfactorical noise and rejected. Odour active components were defined as those which were detected at an identical retention time by at least three panellists. Since the perception and description of odour qualities is individually subjective, a discussion among six panellists is performed subsequent to the sniffing session to get the identical terms of described odours.

5.2.6. Statistical analyses

All data of external and nutritional properties were subjected to a standard analysis of variance (ANOVA), with significant difference ($P \leq 0.05$) between means (Steel et al., 1996) and further analysed with Duncan test using the statistical program SPSS 11.5 for Windows (SPSS Inc., Chicago, USA, 2000). Data of the consumer panel test were subjected to non-parametric Kruskal-Wallis test using the Program MINITAB 14 (Minitab Inc., State College, USA, 2001). Correlation coefficient test by Pearson was carried out using the Program SPSS 11.5 for Windows (SPSS Inc., Chicago, USA, 2000).

5.3. Results and Discussions

5.3.1. External properties

The external properties of salak fruits are summarised in table 20. "Pondoh hitam" fruits were significantly less weight than "pondoh manggala" and "gading" (15 and 13 %, respectively). A greater edible portion was found in fruits of "gading" and "pondoh super" than in "pondoh hitam" and "pondoh manggala". These results are in accordance with Kusumo (1995). Another author reported that "pondoh manggala" fruits had a greater edible portion than "pondoh super" and "pondoh hitam" (Djaafar, 1998). Obviously, growing conditions and orchard management factors such as soil and water availability, fertiliser supply and fruit thinning system at the early stage of fruit development strongly influence growth and development of salak fruit (Thamrin, et al., 1998). Apparently, the share of the edible fruit portion is highly variable among salak cultivars. All cultivars tested had very similar fruit weight, length and diameter as was also described by Djaafar (1998).

With the exception of “pondoh super”, the dry weight of “gading” was significantly higher than that of “pondoh hitam” and “pondoh manggala”. Considerable dry weight variation in salak cultivars has been reported by Sosrodihardjo (1986), Djahjadi (1988) and Djaafar (1998), which may be due to the genetical background of the cultivars (Sosrodihardjo, 1986; Prabawati et al., 1996; Djaafar, 1998), storage time (Suhardjo, et al., 1995; Prabawati et al., 1996) and maturity stage (Sosrodihardjo, 1986; Suhardjo et al., 1995; Prabawati et al., 1996).

Plant pigments, which are responsible for fruit peel and pulp colour, include chlorophylls (green colour), carotenoids (yellow and orange colours) and anthocyanins (red, blue and purple colours) (Kader and Barret, 1996). The colour data of salak fruits are presented in table 21. L*, b* and chroma and hue angle values of “gading” peel were significantly higher as compared to the other cultivars, indicating that the peel colour of “gading” was brighter and deeper yellow. This is in accordance with the Javanese name of this cultivar, which means “yellow colour”. Peel L*, a*, b*, chroma and hue angle values of “pondoh hitam” were lower as compared to the other cultivars, indicating a darker ground colour and a light red peel appearance of this cultivar. L* value of “pondoh hitam” pulp was significantly higher than that of “pondoh manggala”. The a* value of “pondoh super” pulp was significantly higher, reflecting a deeper green colour compared to the others. Pulp hue angle of “gading” was significantly lower than the rest salak cultivars. Whereas pulp b* and chroma of “pondoh super” and “gading” were significantly higher than those of “pondoh hitam” and “pondoh manggala”. This indicated a more intensive yellow colour of fruit pulp of “pondoh super” and “gading” as compared to that of the other cultivars. It seems that peel colour of salak fruits is genetically determined. This can be explained by the fact that “gading” is not closely related to the pondoh cultivars. On the other hand, the more intensive yellow pulp colour of “pondoh super” and “gading” agree with findings of Thamrin et al. (1997) and Djaafar and Mudjisihono (1998). It is assumed that the yellow pigments reflect a higher carotenoid content in “pondoh super” and “gading” as compared to “pondoh hitam” and “pondoh manggala”.

Table 20. External properties of different salak cultivars ("gading", "pondoh super", "pondoh hitam" and "pondoh manggala")

Parameter	Cultivar			
	"gading"	"pondoh super"	"pondoh hitam"	"pondoh manggala"
Fresh weight (g)	63.32 ± 1.29 bc	57.83 ± 2.87 ab	55.27 ± 2.22 a	64.82 ± 1.73 c
Edible portion (%)	70.98 ± 0.64 b	68.45 ± 0.46 b	58.53 ± 0.50 a	58.28 ± 1.95 a
Length (mm)	49.73 ± 0.55 a	48.69 ± 1.15 a	48.03 ± 0.61 a	51.98 ± 0.67 b
Diameter (mm)	45.01 ± 0.41 a	44.93 ± 0.81 a	44.71 ± 1.34 a	46.07 ± 0.85 a
Dry weight (%)	21.26 ± 0.29 c	20.38 ± 0.53 bc	19.63 ± 0.23 ab	18.68 ± 0.27 a
Peel L*-value	56.28 ± 1.32 c	34.24 ± 1.07 b	30.63 ± 0.56 a	34.49 ± 0.67 b
Peel a*-value	10.37 ± 0.53 b	9.66 ± 0.14 b	7.32 ± 0.33 a	10.05 ± 0.05 b
Peel b*-value	29.28 ± 0.77 c	10.95 ± 0.76 b	4.92 ± 0.56 a	10.78 ± 0.22 b
Peel chroma	31.06 ± 0.89 c	14.66 ± 0.54 b	8.83 ± 0.59 a	14.78 ± 0.13 b
Peel hue angle	70.55 ± 0.89 c	47.70 ± 3.47 b	33.55 ± 2.96 a	46.32 ± 1.57 b
Pulp L*-value	78.79 ± 0.66 ab	77.96 ± 0.66 ab	77.48 ± 0.77 a	79.93 ± 0.30 b
Pulp a*-value	-0.81 ± 0.34 a	-2.45 ± 0.16 b	-1.47 ± 0.09 a	-1.25 ± 0.14 a
Pulp b*-value	25.99 ± 0.44 c	25.11 ± 0.13 c	20.32 ± 0.43 b	18.12 ± 0.77 a
Pulp chroma	26.02 ± 0.42 c	25.23 ± 0.14 c	20.38 ± 0.43 b	18.17 ± 0.77 a
Pulp hue angle	91.84 ± 0.78 a	95.60 ± 0.69 b	94.15 ± 0.54 b	94.04 ± 1.10 b
Pulp firmness	56.13 ± 1.02 a	56.72 ± 2.16 a	57.32± 1.91 a	60.01 ± 2.10 a

Values represent means ± SE. Different letters in the same row indicate significant differences by Duncan test ($P \leq 0.05$)

Fruit pulp texture of all salak cultivars was varying from 56 to 60 Shore A-units (table 20). This is in the same range of the texture of ripe apples (56 to 66 Shore A-units) whereas ripe peach fruits are softer, varying from 45 to 57 Shore A-units (unpublished data).

5.3.2. Nutritional valuable properties

Soluble solid content (SSC) represents the overall quantity of soluble components (sugars, organic acids, mineral, salts, pectin, proteins and other minor constituents) that contribute to the refractive index (Seymour et al., 1993). On the other hand, organic acids are important intermediate products of the fruit metabolism. Organic acids are metabolised into many different constituents, including amino acids, which are forming proteins. Titratable acidity (TA), specific organic acids and their relative quantities as well as other factors influence the chemical buffering system inside the cell and affect pH of the fruit tissue. In fruits, mainly sugars and organic acids contribute to the taste (Seymour et al., 1993). TA and SSC/TA ratio act as useful predictors of taste in a considerable number of fruit species (Batten, 1989). Data corresponding to SSC and TA of all salak cultivars are presented in table 21. SSC of salak fruit was high, ranging from 18.7 to 19.1 °Brix. In comparison to SSC of annona species ranged from 10 to 16 °Brix (Paull, 1982), that of kiwi is about 6 °Brix (Harman, 1981) and SSC of pineapple is 9 °Brix (Lodh et al., 1972). On the other hand, TA of salak fruits was varying between 0.26 and 0.76 % and sugar/acid ratio of the fruits was ranging from 24.6 to 73.5. Acid content of salak fruit is comparable to tropical fruits, such as mango (0.2 – 0.5 %) or guava (0.3 – 1 %) (Herrmann, 1987), but its sugar/acid ratio is very high. The results of SSC and TA explained the very sweet taste of salak fruits. SSC of "gading" was similar to that of the other cultivars, however TA of "gading" was up to 3 times higher than in the "pondoh" cultivars. On the other hand, sugar/acid ratio of "gading" was very low as compared to the "pondoh" cultivars. These indicated that "gading" possessed the least sweet taste, as also have been reported by Lestari et al. (2003). Great differences in the sugar/acid ratio of salak fruits of different cultivars have been reported by other authors (Sosrodihardjo, 1986; Santoso, et al., 1996; Djaafar, 1998). Apart from the cultivar, the variability of sugar and acid contents is known to be influenced by external factors such as soil humidity (Purbiati et al., 1993; Thamrin, 1998), leaf prunning (Soleh, et al., 1993), amount of fertilisers (Kusumainderawati and Soleh, 1995), climate and site conditions (Padmosudarso, 2000) and fruit maturity stage at harvest (Sosrodihardjo, 1986; Tjahjadi, 1988; Djaafar and Mudjisihono, 1998).

Table 21. Soluble solid content (SSC), organic acids, mineral and fat content of different salak cultivars ("gading", "pondoh super", "pondoh hitam" and "pondoh manggala")

Parameter	Cultivar			
	"gading"	"pondoh super"	"pondoh hitam"	"pondoh manggala"
SSC (°Brix)	18.7	19.1	19.1	18.6
Titratable acidity (Malic acid) (%)	0.76	0.33	0.26	0.55
Sugar/Acid ratio	24.61	57.88	73.46	33.82
Ca (mg/g DM)	0.84 ± 0.04 a	0.88 ± 0.04 ab	1.07 ± 0.04 c	1.03 ± 0.06 b
K (mg/g DM)	14.12 ± 0.64 c	15.32 ± 0.43 c	12.34 ± 0.22 b	8.70 ± 0.52 a
Mg (mg/g DM)	3.17 ± 0.01 a	3.12 ± 0.06 a	3.03 ± 0.03 a	3.15 ± 0.08 a
P (mg/g DM)	0.34 ± 0.01 c	0.47 ± 0.02 d	0.26 ± 0.01 b	0.19 ± 0.02 a
Fat (% DM)	0.27	0.33	0.42	0.36

Values represent means ± SE. Different letters in the same row indicate significant differences by Duncan test ($P \leq 0.05$)

Important fruit minerals include base-forming elements (Ca, K, Mg) and acid-forming elements (P). Ca is primary associated with the cell wall and high Ca content reduces respiration and ethylene production (Kader and Barret, 1996). Furthermore, Ca is related to a delay of ripening, less physiological disorders and extended storage life of fruits. K is the most abundant mineral found in fruits and it always occurs in combination with organic acids. High K content is often associated with increased acidity and improved colour of fruits (Kader and Barret, 1996). Mg is a component of the chlorophyll molecule, which is responsible for the green colour of fruit peel and pulp. P is a constituent of cytoplasmic and nuclear proteins and plays a major role in carbohydrate metabolism and energy transfer in fruits (Kader and Barret, 1996).

Mineral contents of salak fruits are presented in table 22. Calculated on a fresh weight basis (100 g), salak fruits contained 18 to 21 mg Ca, 163 to 312 mg K, 59 to 67 mg Mg and 4 to 10 mg P. To compare these values with mineral contents of other fleshy fruit species, we used ranges given by Herrmann (2001): 5 to 25 mg Ca, 100 to 300 mg K, 5 to 25 mg Mg and 10 to 40 mg P. Compared to many other fruit species, salak cultivars were rich in Mg, Ca and K, but had a relatively low P content. "Pondoh hitam" had the highest Ca content, whereas "gading" had the lowest. On the other hand, K content of "gading" and "pondoh super" was significantly higher as compared to that of "pondoh hitam" and "pondoh manggala". No significant differences in Mg content of all salak cultivars were found. "Pondoh super" had the highest P content followed by "gading", "pondoh manggala" and "pondoh hitam".

Fat content of different salak cultivars is presented in table 21. The fat content of salak fruits was very low (varying from 0.27 % to 0.42 % DM or from 0.06 % to 0.08 % of fresh edible part) in comparison to most of the fruit species, which comprises of 0.1 to 0.2 % fat in the fresh edible part (Hermann, 2001).

Fruit texture as well as taste and nutritional value are related to the carbohydrate content (Kader and Barret, 1996). Carbohydrates include monosaccharides, disaccharides, polymerised saccharides such as starch and cellulose, sugar alcohols, sugar amines and other derivates such as sugar sulfates. Sucrose (disaccharide), glucose and fructose (monosaccharides) are the primary sugars found in most fruits. Carbohydrate fraction and total sugars in salak fruits as compared to apple, pear and tropical/subtropical fruits (orange, ananas, mango, kaki, kiwi) are presented in table 22. Salak fruits had comparable content of carbohydrate fraction and total sugar than those of other tropical and subtropical fruits. On the other hand, as compared to apple and pear, salak fruits possessed a lower content of fructose, similar content of glucose and higher content of sucrose and total sugar. Fruit carbohydrate content of different salak cultivars is given in table 23. The content of fructose and glucose in "gading" was significantly lower than in the "pondoh" cultivars, with the exception of glucose content in "pondoh super". Sucrose was found to be the predominating sugar in all salak cultivars. Similar results were found for six other salak cultivars from Bali (Indonesia), i.e. "gondok", "biasa", "nangka", "nenas", "putih" and "gulapasir" (Suter, 1988). These results reflected high nutritional value of salak fruits in term of sucrose and total sugar. This might indicate that salak fruits were sweet and had higher amount of sugar source used for synthesis of pectic substances and other cell wall materials as well as for converting to the storage product, starch. The used of sugar during life of the fruit has been explained by Whiting (1970)

Table 22. Carbohydrate fraction and total sugars in salak fruits as compared to apple, pear and tropical/subtropical fruits (on a fresh weight basis)

Fruit	Fructose	Glucose	Sucrose	Total sugar
	%			
Salak	2.2 – 3.6	1.9 – 4.0	6.5 – 9.7	13.4 – 14.4
Apple and Pear[1]	3.8 – 7.3	1.3 – 3.8	1.4 – 4.5	6.5 – 13.2
Tropical/subtropical fruits[1]	1.8 – 8.0	0.5 – 5.7	1.3 – 11.0	8.4 – 21.0

[1] Hermann (2001)

Polysaccharides, such as cellulose and hemicellulose, are compounds of the cell wall, whereas pectins are located in the middle lamella (Herrmann, 2001). The fibres provide a high nutritional value for the consumer, and have beneficial effects in preventing cardiovascular diseases, diverticulosis and colon cancer (Kader and Barret, 1996). Cellulose (ß (1→4)-D-polyglucan), forms the skeletal scaffolding of the wall through the formation of microfibrils of 5 to 15 nm in diameter and being several thousand units long. Hemicellulose consists of rigid, highly branched rod-shaped polymers of neutral sugars, such as xylan, xyloglucan and ß(1→3) or ß(1→4) mixed glucans, which are ~200 nm in length and link with cellulose, pectin and lignin by hydrogen bonding (Jackman and Stanley, 1995). Lignin is a non-saccharide-based structure, intimately formed with and infiltrated through cellulose, a hard and rigid matrix of tremendous strength (Cho et al., 1997). The formation of lignin begins in the primary cell wall or middle lamella, but the highest concentration occurs in secondary cell wall (Cho et al., 1997).

Table 23. Carbohydrate fractions of different salak cultivars (“gading”, “pondoh super”, “pondoh hitam” and “pondoh manggala”)

Component	Cultivar			
	“gading”	“pondoh super“	“pondoh hitam”	“pondoh manggala“
Fructose (mg/g DM)	102.34 ± 0.84 a	117.69 ± 2.85 b	186.24 ± 1.28 c	179.10 ± 5.29 c
Glucose (mg/g DM)	107.48 ± 2.05 a	93.74 ± 5.35 a	203.30 ± 1.91 b	192.76 ± 7.41 b
Sucrose (mg/g DM)	443.10 ± 5.16 b	475.80 ± 20.21b	345.29 ± 4.39 a	347.65 ± 7.30 a
Lignin (% DM)	2.92 ± 0.32	0.54 ± 0.04	0.60 ± 0.18	0.47 ± 0.15
Cellulose (% DM)	3.75 ± 0.14	2.89 ± 0.14	2.80 ± 0.11	2.80 ± 0.01
Hemicellulose (% DM)	0.39 ± 0.31	0.72 ± 0.29	0.53 ± 0.09	0.41 ± 0.22
WSP (mg/g DM)	9.88 ± 0.55 a	15.82 ± 0.27 c	12.77 ± 0.89 b	15.76 ± 0.62 c
EDTA SP (mg/g DM)	8.80 ± 0.55 c	6.80 ± 0.26 b	6.01 ± 0.14 ab	5.50 ± 0.38 a
ISP (mg/g DM)	4.18 ± 0.48 a	5.63 ± 0.29 b	5.23 ± 0.10 ab	5.20 ± 0.54 ab
Tot. pectin (mg/g DM)	22.86 ± 1.56 a	28.25 ± 0.57 b	24.01 ± 1.09 a	26.45 ± 1.21 ab

WSP = Water soluble pectin, EDTA SP = EDTA soluble pectin, ISP = Insoluble pectin, Tot. pectin = Total pectin

Values represent means ± SE. Different letters in the same row indicate significant differences
by Duncan test ($P \leq 0.05$)

Dietary fibres such as pectin, cellulose, hemicellulose and lignin of salak are presented in table 22. Calculated on a fresh weight basis, total dietary fibres in salak fruits (sum of cellulose, hemicellulose, lignin and total pectin) were varying between 1.2 and 2 %. For comparative purposes, total dietary fibre in drupe fruits and strawberry is varying from 1.5 to 2 %, whereas content in other berry fruits is about 3 to 5 % (Herrmann, 2001). The lignin content in "gading" was about six times higher than in all "pondoh" cultivars (table 22), whereas cellulose in "gading" was 30 to 34 % higher than that of the others. On the other hand, hemicellulose in gading was lower than that of the "pondoh" cultivars. The content of WSP in "gading" was significantly lower in relation to the other cultivars tested, but "gading" had the highest content of EDTA SP. Content of ISP and of total pectin in "pondoh super" was significantly higher than in "gading". Total dietary fibre in "gading" was 34 to 48 % higher as compared with that of the "pondoh" cultivars. In terms of dietary fibre, these results reflect a higher nutritional value of "gading" as compared to all "pondoh" cultivars.

5.3.3. Sensory attributes

Sensory attributes of different salak cultivars are summarised in table 23. The rank of the attributes "general appearance", "shape", "external colour" and "internal colour" of "pondoh manggala" was higher than those of the other salak fruits. There were no differences in "texture" and "texture acceptance", indicating that all salak fruits were firm and the texture was accepted. The rank of "taste intensity" of "pondoh super" was lower than that of the other cultivars. No differences in "sweetness", "acidity" and "undesired after taste" among salak cultivars tested were found. The highest "taste acceptance" was found in "pondoh manggala" followed by "pondoh super" and "pondoh hitam", and the lowest value was found in "gading". The sensory tests clearly point out the consumer preference of "pondoh manggala" in terms of appearance and taste, whereas "gading" was not accepted.

Correlation between the external properties and chemical analysis and the corresponding sensory attributes are presented in table 24, 25 and 26. Fruit size was significantly correlated with general appearance, indicating fruit size as an important parameter for the external quality of salak. Firmness was significantly correlated with texture acceptance, which implies that the more firm the fruit is, the more acceptable it ranked. The consumer test revealed that "pondoh manggala" proved the firmest fruit, followed by "pondoh hitam", "pondoh super" and "gading". There was a tendency for a correlation between fructose and taste acceptance, indicating that fructose could be considered as a taste indicator for the consumer. "Pondoh manggala" and "pondoh hitam" had a higher fructose content, followed by "pondoh super" and "gading".

Table 24. Sensory attributes of different salak cultivars evaluated by a consumer panel

Attribute	Cultivar			
	"gading"	"pondoh super"	"pondoh hitam"	"pondoh manggala"
General appearance	10.0 a	7.2 a	8.6 a	16.2 b
Shape	6.6 a	10.8 a	7.9 a	16.7 b
External colour	10.1 a	9.2 a	7.6 a	15.1 b
Internal colour	7.6 a	9.5 a	8.3 a	16.6 b
Easy to peel	7.5 a	7.8 a	12.3 a	14.4 a
Aroma	14.2 b	13.8 b	10.0 b	4.0 a
Texture (firm-soft)	7.5 a	9.7 a	11.9 a	12.9 a
Textural acceptance	11.8 a	7.1 a	10.0 a	13.1 a
Sweetness	7.3 a	9.5 a	10.6 a	14.6 a
Acidity	7.8 a	10.5 a	11.7 a	12.0 a
Taste acceptance	5.2 a	8.4 b	12.0 b	16.4 c
Taste intensity	12.1 b	5.8 a	9.6 b	14.5 b
Undesired after taste	12.7 a	7.0 a	7.7 a	14.6 a

Value represent rank means (calculated statistically from 0 to 5 scale values for each parameter) and different letters in the same row indicate significant different by the Kruskal-Wallis test ($P \leq 0.05$)

Table 25. Correlation between fruit properties and sensoric impression of the consumer panel

Attribute/Compound	General appearance	Texture acceptance	Texture (firm-soft)
Length	0.975*	-	-
Diameter	0.960*	-	-
Peel L*-value	0.074	-	-
Peel a*-value	0.497	-	-
Peel b*-value	0.111	-	-
Peel Chroma	0.133	-	-
Peel Hue angle	0.150	-	-
Pulp L*-value	0.957	-	-
Pulp a*-value	0.519	-	-
Pulp b*- value	-0.605	-	-
Pulp Chroma	-0.609	-	-
Pulp Hue angle	-0.278	-	-
Firmness	-	0.958*	0.363
Water Soluble Pectin	-	0.391	-0.513
Insoluble Pectin	-	0.163	-0.697
Total Pectin	-	0.060	-0.730

* Statistically significant ($P \leq 0.05$)

Table 26. Correlation between chemical components and taste attributes evaluated by consumer panel

Component	Taste acceptance	Taste intensity	Sweetness	Acidity	Undesired after taste
Fructose	0.915	0.726	0.744	0.947	0.147
Glucose	0.813	0.816	0.694	0.840	0.255
Sucrose	-0.771	-0.869	-0.701	-0.776	-0.350
Sugar/Acid Ratio	0.656	-0.095	0.274	0.804	-0.568
SSC	0.003	-0.658	-0.431	0.273	-0.970*

* Statistically significant ($P \leq 0.05$)

5.3.4. Aroma and volatile compounds

The perception of flavour relies on taste and smell. The characteristic flavour of an individual fruit is usually derived via our sense of odour and is due to the production of specific volatiles (Seymour et al., 1993). The main aroma compounds in salak fruits were dominated by carboxylic acids and their methyl esters (table 26). Some alcohols, furanones, and aldehydes were also found among the identified compounds. Some large and medium peaks in the gas chromatogram still remained unknown. The chromatograms of Freon extracts contain 85, 101, 104 and 70 discrete peaks for "gading", "pondoh super", "pondoh hitam" and "pondoh manggala" cultivars, respectively. Constituents of the salak fruit extracts which could be identified by the GC-MS were in accordance with aroma patterns found by Wong and Tie (1993) and Supriyadi et al. (2002).

The presence of a large number of methyl esters of C5 and C6 branched-chain alkanoic, alkenoic and hydroxy alkanoic acids with their isomers have been rarely found in other tropical fruits than salak (MacLeod and deTroconis, 1982; Sakho et al., 1997; Jirovetz et al., 2002; Jordan et al., 2003). The differences of the relative concentration of volatile compounds in the four different salak cultivars are summarised in table 26. In comparison to all "pondoh" cultivars, "gading" might be characterised by a higher relative concentration of methyl 2-methylbutenoate and (*E*)-2-methylbutenoic acid but a lower content of the unknown esters 1 and 2. "Pondoh hitam" and "pondoh manggala" had almost similar relative concentrations in most of the aroma compounds.

The NIF profile (top) versus the FID gas chromatogram (bottom, inverted) of a salak fruits extract is presented in figures 28 and 29. The minor compounds probably play an important role in the flavour composition of salak (Valim et al., 2003; Jordan et al., 2003). Some of the intense olfactory responses were found in regions with small FID signals, i.e. peak 1, 3, 6, 17, 18, 20 and some others. On the other hand, many of the major compounds such as methyl 3-hydroxy-3-butanoate, methyl hexanoate, methyl 3-methyl-hydroxy pentanoate or the big unknown peak produced little or no olfactory response. The use of GCO technique allowed the detection of components that were unable to be detected by FID. For example, peak 8, described as cooked potato aroma, which was usually corresponding to the presence of methional. No peak in the FID chromatogram correlated with this compound. This phenomenon has been previously reported by Valim et al. (2003). Since methional has a very low olfactory threshold of 0.2 ng/l in water (Valim et al., 2003), it was recognised by the panelists even in very low concentration. This small olfactory active peak might be overlapped by the large methyl 3-hydroxy-3-methyl butanoate peak. The GC-MS spectrum indicated that this big nearest peak contains more than one compound. A similar phenomenon was found at the high boiling point area, such as peak 23 and 24 which have a strong and intense olfactory impact although no chromatogram-peaks can be correlated with it. In case of peak 21 and peak 22, the compounds were tentatively identified by stir bar sorptive extraction (SBSE) as well as by the aroma description in comparison with the literature (Acree and Arn, 1997; Anonymous, 2002; Tu et al., 2002). The SBSE extraction was used parallely to the liquid extraction for MS identification.

The main differences between the aroma patterns of “pondoh” and “gading” cultivars could be attributed to peaks 2, 3, 6, 15 and 23, which were perceived by almost all of the panelists in the “pondoh” samples, but could not be detected in “gading”. According to the olfactory results, they contributed to metallic, chemical, rubber-like, strong green, woody, flowery and vanilla like characters in both “pondoh” cultivars. The main differences between the “pondoh” cultivars occurred at peak 2, which contributes to the typical salak aroma. This olfactory response could be clearly detected by panellists for “pondoh super”, but not for “pondoh hitam” and “gading”. The obtained response is related to methyl 2-methylbutanoate. Higher availability of this compound in “pondoh super” might be due to the specific aroma of this cultivar.

Table 27. Relative concentration to internal standard of major volatiles of different salak cultivars ("gading" (G), "pondoh super" (PS), "pondoh hitam" (PH) and "pondoh manggala" (PM))

No	Compound	Cultivar			
		G	PS	PH	PM
1	2,6-Dimethyl-5-heptenol (IS)*	1.00	1.00	1.00	1.00
2	Methyl 2-methyl-butanoate	0.27	2.16	0.60	0.26
3	Methyl 3-methyl-butanoate	0.54	1.86	-	-
4	Methyl 3-methyl-pentanoate	5.01	10.11	2.96	2.27
5	Methyl 3-methyl-2-butenoate	3.01	4.52	0.81	0.27
6	Methyl 2-methyl-butenoate	1.58	0.86	0.34	0.36
7	Methyl 3-methyl-3-pentenoate	0.36	0.48	0.35	0.31
8	Methyl 3-methyl-2-pentenoate	34.42	35.83	14.14	8.72
9	Methoxybenzene	0.28	0.36	0.39	0.28
10	Methyl 3-hydroxy-3-methyl-butanoate	3.53	22.50	11.38	4.04
11	Unknown Ester 1	34.20	82.60	61.08	47.92
12	Unknown Ester 2	6.54	12.18	13.48	8.62
13	4,5-Dimethyl-dihydro-2(3H)-furanone	18.38	15.20	9.11	14.58
14	2-Methyl-butanoic acid	70.79	112.61	75.42	65.02
15	6-Methyl-tetrahydro-2H-pyran-2-one	10.64	23.84	16.64	6.49
16	5-Ethyl-dihydro-2(3H)-furanone	0.31	6.39	1.84	0.28
17	Unknown 1	5.81	10.98	12.44	4.34
18	3-Methyl-pentanoic acid	13.92	14.82	4.23	2.42
19	3-Methyl-butenoic acid	1.31	1.37	0.41	0.21
20	[*E*]-2-Methyl-butenoic acid	6.52	2.99	5.49	3.10
21	[*E*]-2-Methyl-2-pentenoic acid	3.18	5.37	3.28	2.19
22	Phenylethyl alcohol	0.41	0.39	1.05	0.34
23	Unknown 2	6.00	7.01	4.46	3.91
24	Unknown 3	5.64	16.25	7.34	3.04
25	Methyl 4-hydroxy-3-methyl-pent-2-enoate	54.33	76.98	57.28	38.28

Note: (IS)* = Internal standard = 10 μl standard solution per 100 ml sample solution

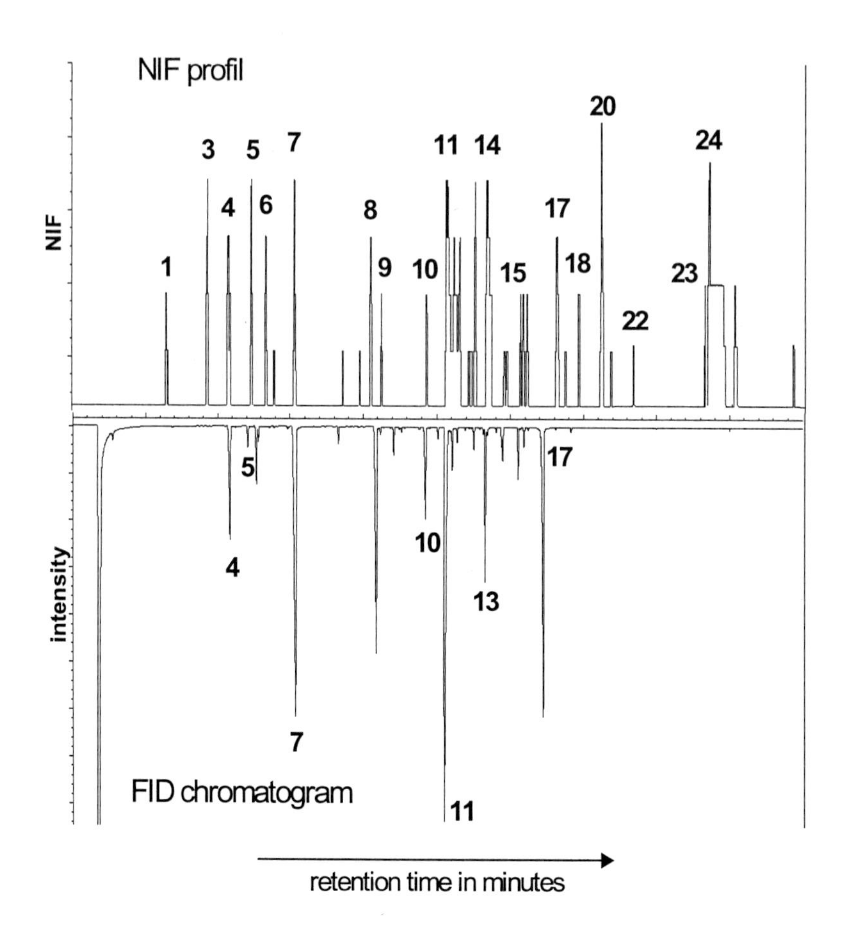

Figure 28. Comparison of NIF profile and FID chromatogram of salak "pondoh hitam" (PH).

The compound numbers refer to table 28

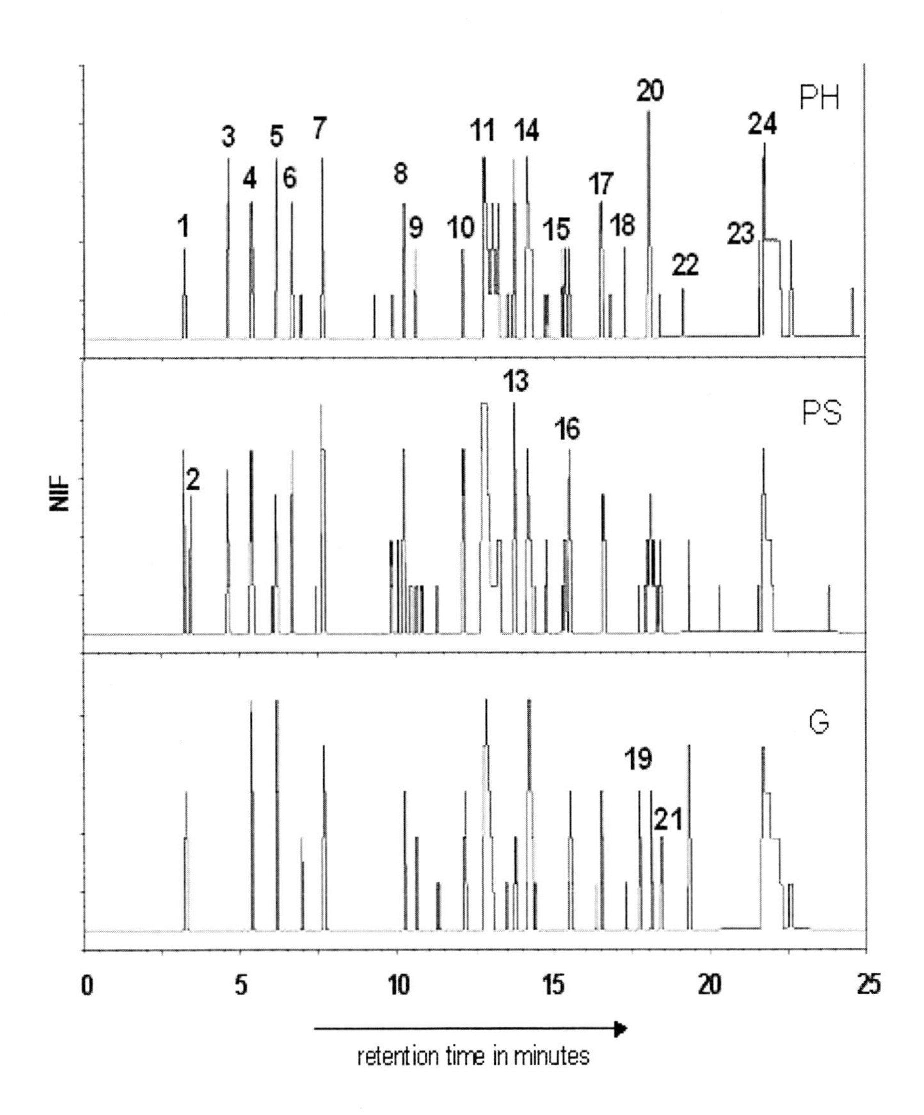

Figure 29. NIF profiles of different salak cultivars ("pondoh hitam" (PH), "pondoh super" (PS) and "gading" (G)).
Odour responses with a nasal impact below NIF = 2 were excluded (olfactory noise).
The compound numbers refer to table 28

Table 28. Results of the GCO analysis of different salak cultivars (“pondoh hitam” (PH), “pondoh super” (PS) and “gading” (G))

No	RT[1] (min)	RI[2]	compound	odor description	Ident.[3]	NIF Cultivar		
						PH	PS	G
1	3.25	1011	Methyl 2-methylbutanoate	fruity, fresh	1, 2, 3	5	4	4
2	3.46	1020	Methyl 3-methylbutanoate	cheesy, typical salak	1, 3	1	4	nd
3	4.63	1085	2-Methyl-2 butenal	metallic, rubbery, green	1, 3	5	5	nd
4	5.34	1127	Methyl 3-methylpentanoate	fruity, sweet, typical salak	1, 2, 3	5	5	6
5	6.16	1184	Methyl 3-methyl-2-butenoate	overripe fruit, fruity, green/ethereal	1, 3	5	5	6
6	6.66	1209	unknown	pyrazine, metallic, chemical, woody, strong green	--	4	5	nd
7	7.63	1262	Methyl 3-methyl-2-pentenoate	ethereal, woody, strong green, spoilaged	1, 2, 3	5	6	5
8	10.24	1463	Methional	cooked potato	2	4	5	4
9	10.59	1478	unknown ester (mz 41, 55, 87)	fruity, sweet, typical salak (weak)	1	3	2	3
10	12.13	1591	(3S, 2R)-3,4-dimethyl- butyrolactone	warm, woody, burnt caramel, sharp spicy	1, 2	3	5	4
11	12.78-13.24	1658	2-Methylbutanoic acid	cheesy, unpleasant overripe, sweaty, sour, buttery	1, 2, 3	5	6	6
12	13.75	1741	Tetrahydro-2-methyl thiophene (mz 87, 102)	stinky flowery, almond, stale	1	5	6	3
13	14.16	1780	3-Methylpentanoic acid, 3-methyl-2-butenoic acid	unpleasant riped cheese, rancid, pungent	1, 3	5	5	6

Table 28. (continued)

#	RT (min)	RI	compound	odor description	Ident.	NIF cultivar		
						PH	PS	G
14	14.80-14.93	1835	2-Methyl-2-butenoic acid	marcipan, unpleasant	1, 2, 3	3	3	1
15	15.28-15.36	1890	Phenylethyl alcohol	flowery, floral, drop acid (honey), fragrant	1, 2, 3	3	3	nd
16	15.54	1909	(*E*)-3-Hexenoic acid	stinky wet cloth, sharp pungent, vinegar, sweaty	1, 2, 3	3	5	4
17	16.54	2020	2,5-Dimethyl-4-hydroxy-3 (2H)-furanone	caramel, sweet, cotton candy	1, 2, 3	4	5	4
18	17.29	2113	unknown	chemical, burnt	--	3	1	3
19	17.71	2215	unknown	spicy, clove, christmas cake	--	nd	2	4
20	18.04	2226	Methyl anthranilate	strong, lovage, celery-leaves, "Maggi"	1, 2, 3	6	5	3
21	18.38	2276	Methyl dihydrojasmonate	flowery, perfumery	2, 4	2	3	3
22	19.27	2378	Isoeugenol	spicy, cooked meals, smoky	2, 4	3	3	5
23	21.54	>2600	unknown	vanilla-like		3	2	nd
24	21.67	>2600	unknown	stinky flowery, silage		5	5	5

[1] Retention time at GCO; [2] Retention index; [3] Substance identification by 1 - MS library search, 2 - aroma descriptors, 3 - reference data, 4: SBSE-MS and library search

The result of GCO analysis is presented in table 28. The unique flavour of salak fruits might be due to these various characters of the olfactory responses. The responses were varying from the desired ones like “flowery”, “fruity”, “sweet”, “caramel” to undesired impressions like “cheesy”, “sweaty”, as well as the light odour like “green”, “fresh”, “cooked potato” to the strong ones like “overripe”, “warm”, “spicy”, “pungent” and “woody”. The significant aroma of fruits is not the result of only one odour impression, but all potential odorants are responsible for the overall flavour impression (Sakho, et al., 1997). Methyl 3-methyl pentanoate, methyl 3-methyl-2-pentenoate and methyl 3-methyl butanoate were found to be the compounds which were predominantly responsible for the salak fruit flavour since most of panellists were able to recognise it with the impression “typical salak“ (Schieberle and Hofmann, 1997; Supriyadi et al., 2002). Methyl dihydrojasmonate and isoeugenol, which also have an olfactometric impact, were identified for the first time as salak fruit volatiles.

In the future, emphasis should be placed to eliminate cultivars with undesired flavour compounds such as 2-methylbutanoic acid (Jezussek et al., 2002; Leudauphin et al., 2003), 3-methylpentanoic acid, 3-hexenoic acid (Supriyadi et al., 2002) and presumably another carboxylic acid leading to the impression “cheesy”, “sweaty”, “overripe fruit” and “butter sour” in order to improve the acceptability of salak on fruit markets outside Indonesia.

5.4. Conclusions

From this study, it is concluded that salak fruits posses a high nutritional value with regard to soluble solid contents, disaccharides and minerals (K, Ca, Mg), but are relatively low in organic acids. With regard to contents of glucose and fructose, salak fruits are comparable to other tropical fruits. In respets to contents of total dietary fibre, salak fruits are similar to some temperate-zone fruits. Genetical factors (i.e. the cultivar) have a strong effect on the nutritional value of salak fruit, since there were significant differences between “gading” and the “pondoh” cultivars, which are not closely related. All “pondoh” cultivars had a higher soluble solid content (SSC), sugar/acid ratio, fructose, hemicellulose and pectic substances (water soluble pectin, insoluble pectin, total pectin) in comparison to “gading”. On the other hand, “pondoh” cultivars had a lower content of lignin, cellulose, EDTA soluble pectin and total dietary fibres than "gading". "Pondoh" fruits were characterised by a sweeter taste, higher content of pectic substances and lower total dietary fibres content as compared to "gading".

High correlations were found between fruit size and general appearance as well as between firmness and texture acceptance. Fructose content and taste acceptance evaluated by consumer panels, were also highly correlated. Fruit characteristics such as weight, size, firmness and fructose content can be used as criteria for the consumer acceptance of salak fruits. "Pondoh manggala" is the most interesting cultivar for the production due to its outstanding consumer acceptance, whereas "gading" seems not to be appropriate for the fresh fruit market. However, the high content of nutritional valuable dietary fibres in "gading" should be considered for the supply of the local people of the production regions.

The unique aroma of salak is the result of a complex composition of volatile compounds, some of them being still unidentified. The major aroma compounds of salak fruits consist of methyl esters of C5 and C6 branched-chain alkanoic, alkenoic and hydroxy alkanoic acids with their isomers. These patterns seem to be unique as compared to other tropical fruit species. Our studies on the aroma composition and consumer acceptance of salak fruits will provide valuable information of fruit quality and for fruit processing lines.

6. PHYSICAL AND PHYSIOLOGICAL CHANGES OF SALAK "PONDOH" FRUITS (*Salacca zalacca* (Gaertn.) Voss) DURING MATURATION AND RIPENING

6.1. Introduction

Since fresh or processed fruit form an important part of the human diet, there is an increasing demand for both improved quality and extended variety of available fruits (Seymour et al., 1993). Commercially, trade is dominated by a relatively small number of fruit species such as grape, banana, citrus and apple. However, western consumers are becoming more aware of exotic fruits and the trade volume of these commodities is increasing rapidly (Seymour et al., 1993). Among exotic fruits, the salak fruit of a spiny palm tree (*S. zalacca* (Gaertn.) Voss) has a high potential as an export crop. Among the various cultivars, "pondoh" is known to reveal superior quality, especially in respect to its sweeter taste without bitter or sour components in comparison with other cultivars, even at early ripening stages.

In postharvest physiology, the terms "mature" and "ripe" are considered to describe different stages of fruit development (Reid, 2002). "Mature" is defined as the point at which a commodity has reached a sufficient stage of development that, after harvesting and postharvest handling, its quality will be at least minimum acceptable for the ultimate consumer (figure 30).

Growth of fruits is usually monitored by measuring attributes such as fruit diameter, volume and fresh or dry weight (Bollard, 1970). The complex chemical fruit composition varies greatly during growth and further development. Destructive sampling methods and complex chemical analyses are required due to the fact that most of the individual compounds failed to be useful for a reliable determination and prediction of maturity and product quality (Reid, 2002). During fruit development, significant changes e.g. in colour (Tucker, 1993), content of sugars, acids and minerals will occur (Bollard, 1970; Tucker, 1993). Texture modifications which cause softening usually also occur during postharvest ripening of fruits (Reid, 2002). It has been well established that texture changes in fruits are consequence of modifications undergone by component polysaccharides that, in turn, give rise to disassembly of primary cell wall and middle lamella structures (Jackman and Stanley, 1995).

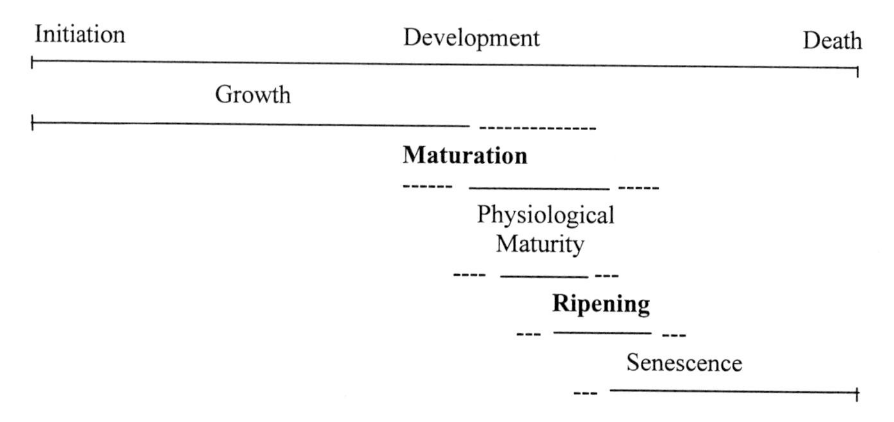

Figure 30. "Maturation" and "ripening" in relation to developmental stages of a fruit (Watada et al., 1984)

Recommendations are required for determining the optimal developmental stage of salak fruits at harvest in respect to nutritional and sensory attributes as well as to postharvest and storage concerns. Increasing the knowledge about the fruit is prerequisite for optimising the consumer-oriented fruit quality and for extending the marketing period, especially for the export market. Therefore, the aim of the present study was to determine physical and physiological changes during fruit maturation and ripening of two salak cultivars, "pondoh hitam" and "pondoh super".

6.2. Materials and Methods

6.2.1. Plant material

For this study, fresh salak fruits of the cultivars "pondoh hitam" and "pondoh super" at three different ripening stages were used, i.e. stage 4, stage 5 and stage 5.5 (4, 5 and 5.5 months after pollination, respectively) for "pondoh hitam" and stage 4, stage 5 and stage 6 (4, 5 and 6 months after pollination, respectively) for "pondoh super". The salak fruits used for the study were grown in the Sleman district, Yogyakarta (Indonesia) and were harvested in September 2003. Immediately after air transport from Indonesia to Germany, the physical measurements and chemical analyses were carried out in Berlin. Freeze-dried sample material was used for the determination of minerals, mono- and disaccharides, pectic substances and dietary fibre. The results are expressed on a dry matter (DM) basis.

6.2.2. Fruit properties

The fruit (n = 7) of each ripening stage (length and diameter) was measured using callipers. Fresh fruits were weighed and the edible portion was determined. Dry weight of each ripening stage was determined using the method as described in Section 5.2.2.

6.2.3. Colour

Measurements of the fruit peel and pulp colour were conducted using a Minolta Colorimeter (CR-321, Minolta, Japan) as described in Section 3.2.4. Colour was recorded on each of the fruits per cultivar. Peel colour was recorded on the equator region of each fruit with eight measurements. Peel colour was computed on the tip and base regions of each fruit with four measurements.

6.2.4. Firmness

The pulp firmness was measured non-destructively using a Shore A instrument (HHP-2001, Bareiss, Germany) as described in Section 5.2.2. The measurement for each fruit was carried out on 3 equidistant records of the equatorial region. The average value per treatment was computed on 7 fruits per cultivar and was expressed as Shore A units.

6.2.5. Soluble solid content and titratable acidity

Measurements of soluble solid content (SSC) were performed on extracted juice of 8 fruits for each ripening stage with two replicates. SSC, titratable acidity and sugar/acid ratio were measured using the methods as described in Section 5.2.3.1.

6.2.6. Minerals

Freeze-dried fruit samples (500 mg), triplicates per each ripening stage, were used for the determination of K, Ca, Mg and P. The analysis was conducted as described in Section 3.2.5. The data of the minerals were expressed as mg/g DM.

6.2.7. Mono- and disaccharides

The analyses were performed in triplicates per ripening stage as described in Section 5.2.3.4. and the data were expressed in mg/g DM.

6.2.8. Pectic substances

Freeze-dried fruit samples (1.5 g) were used for the determination of pectin. The analyses were performed in triplicates per ripening stage as described in Section 5.2.3.6. and the data were expressed in mg/g DM.

6.2.9. Cellulose, hemicellulose and lignin

One g of freeze-dried fruit sample was used for the determination of cellulose, hemicellulose and lignin. The analyses were conducted using the method as described in Section 5.2.3.5. and the data were expressed in % DM.

6.2.10. Statistical analyses

All data were subjected to the standard analysis of variance (ANOVA), with significant differences between means determined ($P \leq 0.05$) (Steel et al., 1996) and then further analysed with Duncan test using the statistic program SPSS 11.5 for Windows (SPSS Inc., Chicago, USA, 2000).

6.3. Results and Discussion

6.3.1. Fruit properties

Physical changes of salak "pondoh hitam" and "pondoh super" during ripening are summarised in table 28. The fruit size of both salak cultivars increased significantly during the ripening process. This result is in agreement to Sosrodihardjo (1986). Fresh weight (FW) of the fruits increased significantly during ripening from stage 4 to stage 5.5 for "pondoh hitam" by 170.8 % and from stage 4 to stage 6 for "pondoh super" by 266.9 %. The edible portion of both cultivars increased until stage 5 and remained constant thereafter. These findings are consistent with findings of Djaafar and Mudjisihono (1998), who reported that fruit weight and edible portion of salak "pondoh hitam" and "pondoh super" cultivars increased until stage 5 and then remained constant until stage 6. Supriyadi et al. (2002) reported that fruit weight and edible portion of salak pondoh increased during maturation. Dry weight of salak fruits increased during ripening by 11 % from stage 4 to stage 5.5 for "pondoh hitam" and by 5 % from stage 4 to stage 6 for "pondoh super". This indicated that the fruit growth in terms of volume increase continuously during maturation and ripening period.

6.3.2. Colour

Colour change of the fruit peel and pulp is associated with ripening and represents a key attribute for the determination of quality (Seymour et al., 1993). Plant pigments, which are responsible for fruit peel and pulp colouration, include chlorophylls (green), carotenoids (yellow and orange) and anthocyanins (red, blue and purple) (Kader and

Barret, 1996). Carotenoid pigments, such as ß-carotene and lycopene are described to be responsible to the colour expression (Seymour et al., 1993).

The peel colour values (L*, a*, b*, chroma and hue angle) of both salak cultivars increased significantly during ripening (table 28), indicating lighter peel with more red and yellow components. It is suggested from our study that colour change of the peel from dark brown to light brown can be used for the determination of fruit ripening stages. Colour values of "pondoh super" were higher at all developmental stages in comparison to those of "pondoh hitam". These reflected a darker peel with less red and yellow colour components of "pondoh hitam" in comparison to "pondoh super". Similar changes in the peel colour of salak during ripening were reported by Djaafar and Mudjisihono (1998), who found that peel colour changed from blackish brown to yellowish brown for "pondoh super" and from blackish brown to black for "pondoh hitam".

L* and a* values of the pulp of "pondoh hitam" declined during ripening, whereas b* and chroma remained constant and hue angle increased. This indicated an acceleration in the development of green compounds, but no changes in yellow pulp colour during ripening. On the other hand, L* of "pondoh super" pulp remained constant and a* declined during ripening, whereas b*, chroma and hue angle values increased significantly. This indicated a faster acceleration of green and yellow pulp colour compounds during ripening in "pondoh super"as compared to those in "pondoh hitam". Pulp colour changes of both cultivars showed that there was no degradation of chlorophylls occurred. According to Tucker (1993) the degradation of chlorophylls could in turn unmask previously present pigments, particularly ß-carotene, which reflected a colour range from yellow to orange. Eventhough, as reported by Setiawan et al. (2001), salak fruit was found to be an excellent source of provitamin A, i.e. containing 1130 µg lycopene/100 g FW and 2997 µg ß-carotene/100 g FW. Most of the colour values of "pondoh super" were higher at all ripening stages in comparison to those of "pondoh hitam", assumingly due to a progressed ripening pattern of "pondoh super". Similar changes in the pulp colour of salak during ripening were reported by Djaafar and Mudjisihono (1998), who found that the pulp colour of "pondoh super" was more yellow in comparison to "pondoh hitam" with a whitish pulp.

Table 29. Quality attributes of salak fruits at different ripening stages

Quality attribute	Salak "pondoh hitam"			Salak "pondoh super"		
	stage 4	stage 5	stage 5.5	stage 4	stage 5	stage 6
Fruit length (mm)	47.05 ± 1.26 a	53.50 ± 1.27 b	53.80 ± 1.21 b	40.84 ± 1.66 a	60.47 ± 2.76 b	70.50 ± 1.26 c
Fruit diameter (mm)	30.14 ± 0.56 a	39.00 ± 0.34 b	45.43 ± 0.73 c	32.21 ± 0.76 a	42.31 ± 1.15 b	52.84 ± 0.70 c
Fresh weight (g)	17.87 ± 0.60 a	34.59 ± 0.92 b	48.40 ± 1.04 c	20.17 ± 0.68 a	39.94 ± 1.79 b	74.00 ± 3.47 c
Edible portion (mg/g)	517.10 ± 19.9 a	592.00 ± 23.9 b	570.20 ± 11.4 ab	365.80 ± 49.4 a	665.90 ± 58.0 b	674.20 ± 18.7 b
Dry weight (mg/g)	165.55 ± 0.80 a	183.40 ± 9.00 b	186.20 ± 1.40 b	177.90 ± 12.4 a	195.90 ± 2.50 a	186.20 ± 9.50 a
Peel L*-value	29.15 ± 0.25 a	29.87 ± 0.16 a	30.96 ± 0.13 b	30.34 ± 0.37 a	33.16 ± 0.59 b	33.59 ± 0.48 b
Peel a*-value	4.87 ± 0.25 a	5.24 ± 0.17 a	7.51 ± 0.21 b	7.66 ± 0.33 a	10.19 ± 0.43 b	10.14 ± 0.37 b
Peel b*-value	3.02 ± 0.24 a	3.50 ± 0.16 a	5.21 ± 0.24 b	6.50 ± 0.36 a	10.34 ± 0.93 b	10.50 ± 0.78 b
Peel chroma	5.73 ± 0.33 a	6.30 ± 0.23 a	9.14 ± 0.30 b	10.05 ± 0.46 a	14.55 ± 0.94 b	14.61 ± 0.80 b
Peel hue angle	31.56 ± 2.54 b	33.66 ± 0.42 ab	34.67 ± 0.79 a	40.22 ± 0.98 b	44.80 ± 1.75 a	45.69 ± 1.19 a
Pulp L*-value	76.61 ± 0.41 b	77.30 ± 0.54 b	74.80 ± 0.66 a	78.86 ± 0.87 a	78.15 ± 0.43 a	78.93 ± 0.46 a
Pulp a*-value	0.60 ± 0.23 b	-0.43 ± 0.16 a	-0.25 ± 0.29 a	0.47 ± 0.19 b	-1.08 ± 0.07 a	-1.29 ± 0.13 a
Pulp b*-value	16.86 ± 1.12 b	19.69 ± 0.44 a	17.04 ± 1.27 b	15.52 ± 0.69 a	24.55 ± 0.37 b	26.77 ± 0.35 c
Pulp chroma	16.89 ± 1.13 a	19.70 ± 0.44 a	17.05 ± 1.27 a	15.52 ± 0.69 a	24.57 ± 0.37 b	26.80 ± 0.35 c
Pulp hue angle	87.95 ± 0.75 b	91.28 ± 0.43 a	90.87 ± 0.30 a	88.40 ± 0.65 b	92.69 ± 0.21 a	92.77 ± 0.28 a
Pulp firmness	51.59 ± 1.77 a	55.31 ± 1.24 a	54.10 ± 1.13 a	50.25 ± 4.13 a	53.29 ± 1.14 a	46.30 ± 2.22 a

Values represent means ± SE. Different letters in the same line indicate significant differences by Duncan test ($P<0.05$)

6.3.3. Firmness

There were no significant differences in fruit pulp firmness of "pondoh hitam" and "pondoh super" during ripening, varying from 51.6 to 55.3 Shore A units for "pondoh hitam" and from 46.3 to 53.3 for "pondoh super", respectively (table 29). These values reflected a very firm fruit pulp during the entire maturation process. This result was in contrast with the finding of Supriyadi et al. (2002) that the firmness in pulp of "pondoh" fruits increased until stage 5.5, but declined thereafter to the end of ripening period. These different results were possibly due to the different firmness measurement method applied. Supriyadi et al. (2002) measured the texture of a piece of salak pulp 1 x 1 cm^2 in size. On the other hand, the non destructive method applied in this study that involves compression and tension might not be sensitive enough for texture measurement of salak fruit pulp, which contained 1 to 3 hard kernels. At the last ripening stage, "pondoh super" was significantly softer than "pondoh hitam". This might indicate a faster break down of pectic substances in the middle lamella which could lead to more limitation in shelf life of "pondoh super" as compared to that of "pondoh hitam".

6.3.4. Soluble solid content and titratable acidity

Fruits differ in their relative contents of sugar and acids (Ulrich, 1970; Whiting, 1970). Changes in the chemical composition of salak fruits during ripening are shown in table 29. Soluble solid contents (SSC) of "pondoh hitam" remained constant during the ripening period (varying from 18.5 to 18.9 °Brix). In contrast, SSC of "pondoh super" at stage 4 was relatively low (11.8 °Brix), but markedly increased (55 %) until stage 5 and declined thereafter by only 8 %. The increase of sugar during ripening was possibly due either sugar import from the leaf or to the break-down of starch reserves of the fruit (Seymour et al., 1993). Titratable acidity (TA) of "pondoh hitam" declined gradually (29 %) from stage 4 to stage 5.5, whereas that of "pondoh super" showed a strong decline of 81 % from stage 4 to stage 6. Similar results were reported by Djaafar and Mudjisihono (1998), who found that acidity of both salak cultivars declined during ripening. The decline of acids during ripening, presumably due to their utilisation as respiratory substrate, has been explained by Ulrich (1970). The sugar/acid ratio of "pondoh super" strongly increased from 13.1 (stage 4) to 99.4 (stage 6), while this ratio in "pondoh hitam" increased only slightly, i.e. from 48.7 (stage 4) to 69.3 (stage 5.5). These results showed that respiratory processes of "pondoh super" were on a higher level than that of "pondoh hitam" during ripening, which might indicate a different climacteric ripening patterns of "pondoh super" and "pondoh hitam".

Table 30. Soluble solid content (SSC) and organic acids of salak fruits at different ripening stages

Index	Salak "pondoh hitam"			Salak "pondoh super"		
	stage 4	stage 5	stage 5.5	stage 4	stage 5	stage 6
SSC (°Brix)	18.5	18.9	18.7	11.8	18.3	16.9
Titratable acidity (malic acid) (%)	0.38	0.34	0.27	0.90	0.43	0.17
Sugar/acid ratio	48.68	55.59	69.26	13.11	42.56	99.41

6.3.5. Minerals

According to Bollard (1970) developing fruits may receive nutrients "directly" from the environment or nutrients enter the developing fruit "indirectly". Direct uptake means that nutrients are obtained either from the roots via the xylem stream and from soluble carbohydrates of the leaves via the phloem. In contrast to that, nutrients supplied to the leaves via the xylem stream and then recirculated to developing fruits through the phloem are regarded as absorbed "indirectly". In salak "pondoh hitam", K, Ca and Mg declined during ripening, whereas P content remained relatively constant (table 30).

These results are consistent with Imad et al. (1995), who reported a decline of K, Ca and Mg during ripening of date fruit. This may be caused by the active competition between the developing fruits and the growing leaves and shoots as it has been reported for grapes (Combe, 1962), Japanese pear (Buwalda and Meekings, 1990) and sapodilla (Sulladmath, 1983).

Table 31. Mineral content of salak fruits at different ripening stages

Element	Salak "pondoh hitam"			Salak "pondoh super"		
	Mineral content (mg/g DM)					
	stage 4	stage 5	stage 5.5	stage 4	stage 5	stage 6
K	18.29 ± 0.26 b	12.66 ± 0.37 a	12.59 ± 0.56 a	20.28 ± 0.33 c	10.65 ± 0.04 b	12.52 ± 0.37 a
Ca	1.17 ± 0.00 c	0.85 ± 0.01 b	0.55 ± 0.03 a	1.00 ± 0.02 c	0.42 ± 0.02 b	0.26 ± 0.00 a
Mg	0.79 ± 0.00 c	0.60 ± 0.01 b	0.54 ± 0.02 a	0.86 ± 0.01 c	0.61 ± 0.01 b	0.58 ± 0.01 a
P	0.75 ± 0.00 a	0.74 ± 0.01 a	0.79 ± 0.03 a	1.15 ± 0.01 b	1.05 ± 0.04 a	0.90 ± 0.03 a

Values represent means ± SE. Different letters in the same line indicate significant differences by Duncan test ($P<0.05$)

6.3.6. Mono- and disaccharides

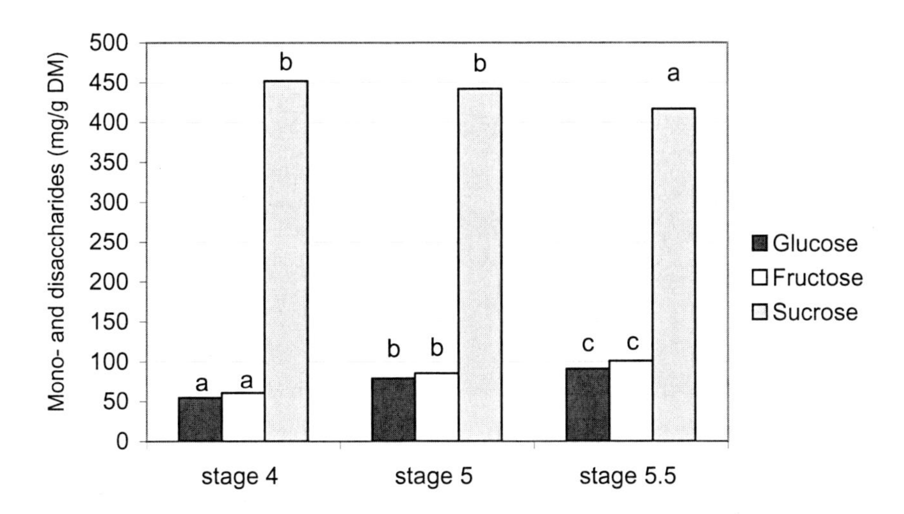

Figure 31. Mono- and disaccharides of salak "pondoh hitam" fruits at different ripening stages. Different letters in the same parameter indicate significant differences by Duncan test (P<0.05)

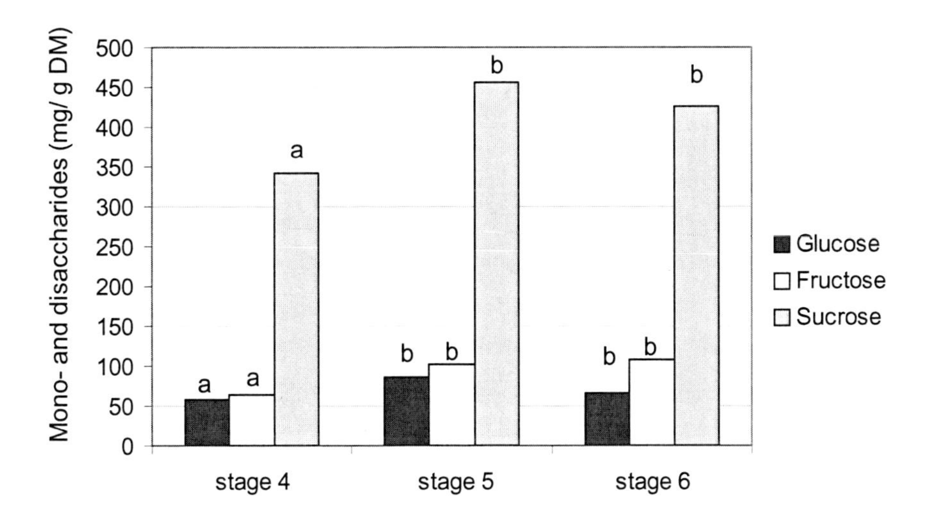

Figure 32. Mono- and disaccharides of salak "pondoh super" fruits at different ripening stages. Different letters in the same parameter indicate significant differences by Duncan test (P<0.05)

During ripening, the contents of glucose and fructose in "pondoh hitam" increased significantly by 67 % from stage 4 to stage 5.5, whereas the content of sucrose remained constant until stage 5, and declined by 5.7 % thereafter (figure 31). On the other hand, the content of glucose, fructose and sucrose in "pondoh super" increased from stage 4 to stage 5 by 50 %, 41 % and 33 %, respectively. The content of glucose and sucrose decreased by 24 % and 6.5 % thereafter, whereas the fructose content remained constant (figure 32). The decrease of sucrose content in "podoh hitam" and "pondoh super" fruits after stage 5 indicated the occurrence of the hydrolysis of sucrose by sucrase to yield glucose and fructose, which reflected accelerated ripening processes in the fruit. This result was in agreement with Supriyadi et al. (2002) and Lestari et al. (2004). The increase of glucose and fructose content in "pondoh hitam" especially after stage 5 might also due to sucrose hydrolisis process. On the other hand, the decrease of glucose content and the constant content of fructose in "pondoh super" after stage 5 might cause by the usage of these monosaccharides, mainly glucose as primary respiration substrates. This possibly indicated a later but faster ripening process and less shelf life in "pondoh super" than in "pondoh hitam" fruits.

6.3.7. Pectic substances

The breakdown of pectic substances and hemicelluloses in the middle lamella weakens the cell wall and reduces the cohesive binding forces between cells. This is strongly associated with the reduction of the fruit firmness (Wills et al., 1981). During ripening, fruits undergo a softening, which is a major quality attribute that often limits shelf life (Tucker, 1993). As ripening progresses, insoluble pectin in the cell wall is converted into soluble pectin by the action of cell wall hydrolysis, which is indicated by the decrease in the content of insoluble pectin (ISP) and the increase of water soluble pectin (WSP). During ripening of salak "pondoh hitam", the ratio of WSP to ISP remained constant until stage 5 and increased by 12 % thereafter. This was caused by the increase of WSP and ISP contents in "pondoh hitam" until stage 5 (figure 33). On the other hand, the ratio of WSP to ISP of "pondoh super" increased more than twice from ripening stage 4 to stage 5 and increased thereafter by 17 % at stage 6. This was due to the increase of WSP in "pondoh super" until stage 5, whereas ISP content in this cultivar decreased during ripening (figure 34). On the other hand, EDTA soluble pectin of "pondoh hitam" decreased until stage 5 and remained constant thereafter, whereas that of "pondoh super" decreased continuously from stage 4 to stage 6 (figure 33 and 34). These results reflected the faster breakdown of pectic substances in "pondoh super" than in "pondoh hitam" during ripening process, assuming the less self life in "pondoh super" than in "pondoh hitam fruits.

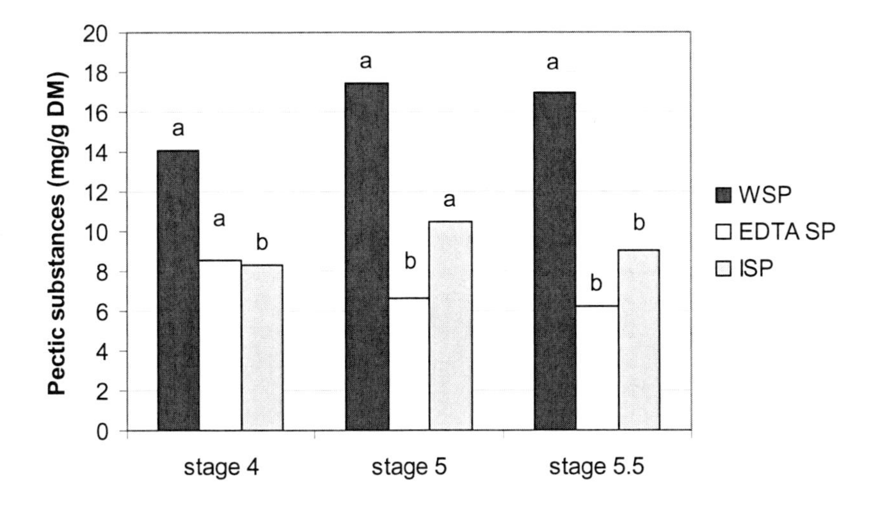

Figure 33. Pectic substances of salak "pondoh hitam" fruits at different ripening stages
WSP : Water Soluble Pectin, EDTA SP: EDTA Soluble Pectin, ISP : Insoluble Pectin
Different letters in the same parameter indicate significant differences by Duncan test ($P<0.05$)

Differences in the change patterns of pectic substances might be due to different activities of pectic enzymes between salak cultivars during ripening stages, such as pectinmethylesterase (PME), polygalacturonase (PG) and cellulase. In some studies, PME activity in tomatoes and bananas during ripening can decline or remain constant or increase depending on the cultivar. On the other hand, PG and cellulase activities appear only with the onset of ripening and tend to increase dramatically during ripening (Seymour, 1997). The monomeric building block of pectin`s polymer backbone consists of α-D-galacturonic acid and methyl α-D-galacturonic acid along with rhamnose units (Cho et al., 1997). The remainder of the structure is made up side chains of the neutral sugars arabinose, galactose, glucose and xylose. PME acts to remove the methyl group from C-6 position of a galacturonic acid, PG hydrolyses the α(1-4) link between adjacent demethyled galacturonic acid residues, whereas cellulase hydrolyses the β(1-4) link between adjacent glucose residues (Seymour et al., 1993). In respect to content of pectic substances of salak fruits during ripening, the activities of PME, PG and cellulase in "pondoh super" might be faster in comparison to those in "pondoh hitam".

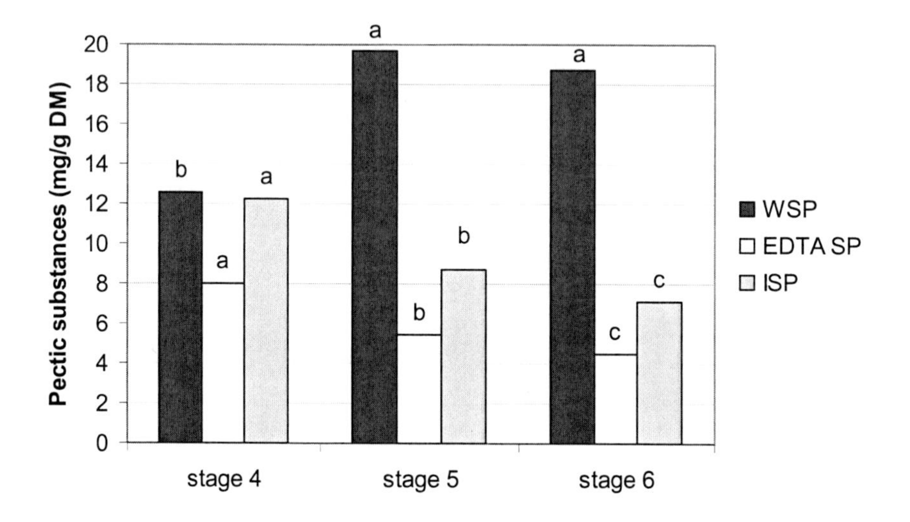

Figure 34. Pectic substances of salak "pondoh super" fruits at different ripening stages
WSP : Water Soluble Pectin, EDTA SP: EDTA Soluble Pectin, ISP : Insoluble Pectin
Different letters in the same parameter indicate significant differences by Duncan test ($P<0.05$)

6.3.8. Cellulose, hemicellulose and lignin

Salak "pondoh hitam" fruits had less cellulose (by 23 %), hemicellulose (by 43 %) and lignin (by 39 %) in comparison to "pondoh super" at ripening stage 4 (figure 35 and figure 36). However, a higher reduction of cellulose, hemicellulose and lignin of "pondoh super" occurred at stage 5 (by 45 %, 64 % and 41 %, respectively) in comparison to "pondoh hitam" at stage 5 (by 20 %, 27 % and 30 %, respectively). These structural carbohydrates content in both salak cultivars decreased continuously thereafter (figure 35 and figure 36). In a study of Manrique and Lajolo (2002), cellulose residues of papaya fruit exhibited decreasing quantities of galacturonic acid and non-glucose monosaccharides during ripening, indicating that an association between polysaccharides from the matrix and microfibrilar phases may be involved in the softening process. Another study gave evidence for the depolymerisation of hemicellulose in tomato during ripening (Huber, 1983). On the other hand, Gross and Walner (1979) reported that the levels of sugars associated with cellulose and hemicellulose in tomato were found to be constant throughout the ripening process. In pears, no correlation between flesh firmness and cellulose content was found and only a slight difference in cellulose and hemicellulose content occurred between hard and soft fruit (Murayama et al., 2002). Depolymerisation of the hemicellulosic fraction in papaya fruit was not evident during ripening (Manrique and Lajolo, 2002).

Cellulose, hemicellulose and lignin are regarded to play a major role in the texture of plant tissue. Instantaneous elasticity of tissue is attributed to the combination of cell turgor and primary cell wall strength, as dictated by cellulose. Viscoelastic properties are related to hemicellulose and pectic components and steady state viscous behaviour to exosmosis and to increased wall fluidity arising from the breakdown of cell wall and/or middle-lamellar polymers (Jackman and Stanley, 1995). The different amount in the declines of structural carbohydrates between two salak cultivars tested in this study might imply a faster break-down of the fruit cell wall in “pondoh super” in comparison to “pondoh hitam” during ripening. This result might also reflect less shelf life in “pondoh super” than in “pondoh hitam” fruits.

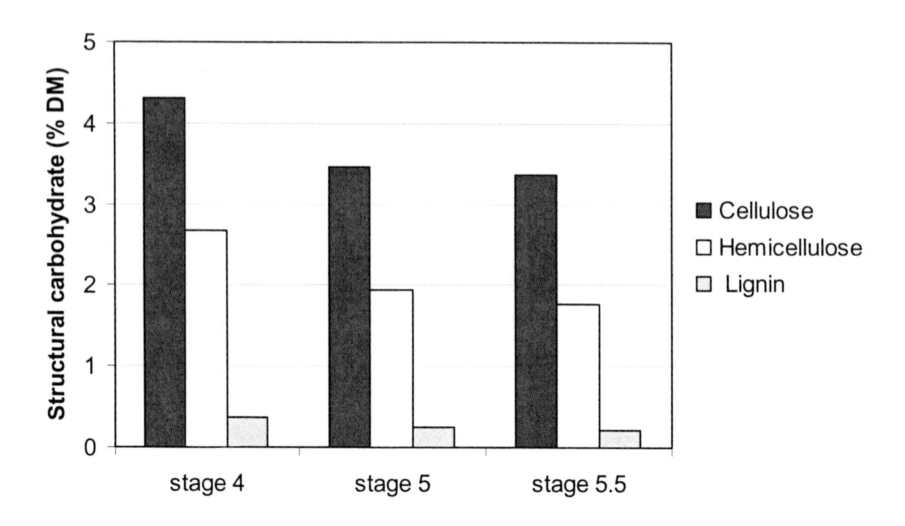

Figure 35. Structural carbohydrate of salak "pondoh hitam" fruits at different ripening stages

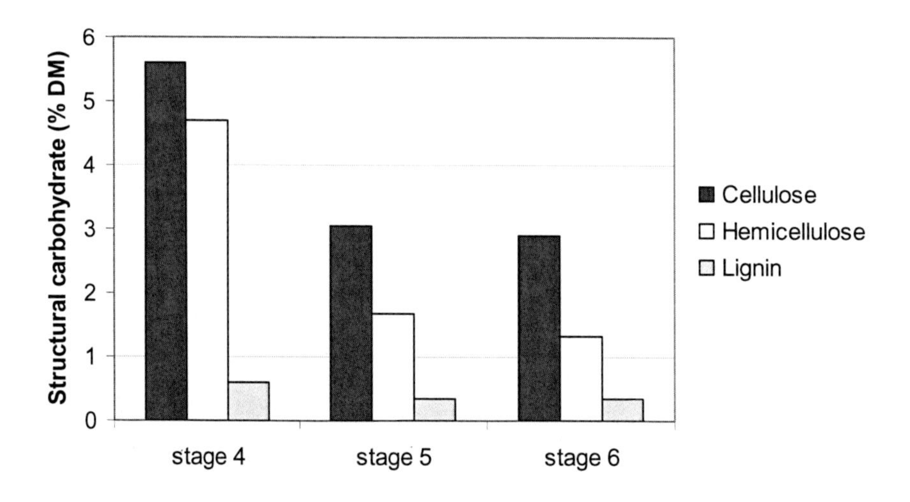

Figure 36. Structural carbohydrate of salak "pondoh super" fruits at different ripening stages

6.3.9. Firmness and relation to pectic substances

There was no significant difference in pulp firmness between "pondoh hitam" and "pondoh super" at all ripening stages. However, there was a marked loss of cell wall compounds during ripening (figures 33 and 34). These results indicated that the method for determining textural properties of salak fruit had not been sensitive enough. Similar findings were also reported for tomato fruit (Hobson and Grierson, 1993). Non-destructive methods that involve compression and tension can result in a misleading texture evaluation, which should be measured objectively in terms of force, distance and time (Jackman and Stanley, 1995). Another possible explanation might relate to the specific enzyme activities in salak fruits. Giovannoni et al. (1989) reported the over-expression of polygalacturonase in non-softening mutants of transgenic *rin* (ripening inhibitor) tomato fruits failed to induce softening, although pectin depolymerisation and solubilisation occurred. Tieman and Handa (1994) noticed a marked loss in tissue integrity during senescence but little modification of transgenic tomato fruit firmness during ripening, which contained an antisense pectinmethylesterase gene, whose expression leads to a 10-fold reduced enzyme activity.

6.4. Conclusions

Increase in fruit size and weight as well as changes in peel and pulp colour occured during maturation and ripening of salak fruits. Different patterns of peel and pulp colour changes were found in "pondoh super" and "pondoh hitam" during ripening.

Physiological processes in "pondoh super" occurred to a later stage but then accelerated faster than "pondoh hitam" in term of changes of mono- and disaccharides, resulting in a poorer marketability and shorter shelf life. In respect to the change of sugar/acid ratio, there was a faster ripening process in "pondoh super" than in "pondoh hitam". "Pondoh super" possessed higher content of polysaccharides and lignin, however, the ripening process accelerated earlier in comparison to "pondoh hitam". Alterations in cell wall and middle lamella structure were not associated with the physical texture measurement during ripening of salak, assumingly due to the non sensitive method applied.

Fruits of salak "pondoh hitam" and "pondoh super" should be harvested from 4 to 6 months after pollination, depending on the market orientation. Thus, if the target is to achieve optimal sensory attributes and nutritional contents, harvest time could be delayed in order to improve those quality compounds. Fruits at stage 6 reached a maximum amount of flavour component, but had lower mineral content and structural carbohydrates in comparison to fruits of an early ripening stage. On the other hand, if the mineral and structural carbohydrate contents are the predominant criterion for the market, i.e. for processing purposes, salak fruits should be harvested at an earlier ripening stage (stage 4).

7. GENERAL CONCLUSIONS AND FUTURE PERSPECTIVES

Salak cultivars, selected for superior fruit quality are very important for the Indonesian market and have a high potential for the export markets. As compared to other cultivars, "pondoh" is the most promising one due to its superior fruit quality. Interdisciplinary approach in improving salak production and postharvest technologies has been carried out in Indonesia. The purpose of these activities was to support the development of intensive salak production and to assure high fruit quality.

The present study was focussing on the ecophysiology of salak palm seedlings as well as on postharvest quality properties of different salak fruit cultivars to accomplish the on-going investigations.

Growth and physiological responses of different salak cultivars to different plant growing media were investigated. Growth parameters of four salak cultivars, "pondoh super", "pondoh hitam", "pondoh manggala" and "gading" were tested on three different growing media, i.e. peat, sand and sand/compost mixture. From our results, it was concluded that "pondoh hitam" is the most vigorous cultivar tested growing on different media. In general, the best growing media for salak seedlings growth is sand supplied with a complete nutrition solution. Peat with a very low pH was not suitable as growing medium for salak seedlings due to inhibitory growth effects on the seedlings. Net CO_2 assimilation rate (P_N) of salak seedlings was relatively low as compared to other tropical tree species. Among the cultivars tested, "gading" exhibited the higher P_N. Foliar mineral contents (N, P, K, Ca, Mg) of salak seedlings were within the optimal range for growth in all media tested.

The second ecophysiological study was carried out to investigate the effects of light and water supply on growth and physiological responses of salak seedlings. From the study it is concluded that shading, which reduced sunlight by 70 % was not beneficial for salak seedlings, due to lower growth rate, P_N and N content as compared with plants in non-shading conditions. Six weeks after the treatment, photoinhibitory responses of non-shaded plants were detected, which caused bleaching in salak leaves. Shading conditions to a certain extent are needed for raising salak seedlings under tropical conditions. However, light intensities of less than 250 µmol $m^{-2}s^{-1}$ (PAR) may lead to retarded growth, decreased P_N and inhibited nitrogen uptake. Light intensities above 800 µmol $m^{-2}s^{-1}$ (PAR) possibly caused leaf bleaching and should be avoided. Salak plants did not tolerate drought conditions. However, different water supply did not affect P_N and mineral uptake (N, P, K, Ca, Mg) of salak plants. Light had a stronger effect on P_N and nutrient uptake of salak seedlings in comparison to water supply.

The study on the postharvest quality of salak was carried out to determine and compare the fruit quality of the different salak cultivars with special emphasis on external properties, nutritional values and sensory attributes. It is concluded that salak fruits had a higher nutritional value with regard to soluble solid contents, disaccharides and minerals (K, Ca, Mg), but they contain relatively small amounts of organic acids in comparison to some other tropical or temperate-zone fruits. In respect to glucose and fructose contents, salak fruits are comparable to other tropical fruits. Regarding total dietary fibre, salak fruits are similar to some temperate-zone fruits. Genetical factors (i.e. the cultivar) have a strong effect on the nutritional value of salak fruit. Since "gading" is not closely related to the "pondoh" cultivars, the two groups of cultivars had significant differences in their nutritional compounds. All "pondoh" cultivars had a higher soluble solid content (SSC), sugar/acid ratio, fructose, hemicellulose and pectic substances (water soluble pectin, insoluble pectin, total pectin) in comparison to "gading". On the other hand, "pondoh" cultivars had a lower content of lignin, cellulose, EDTA soluble pectin and total dietary fibres than "gading". "Pondoh" fruits were characterised by a sweeter taste, higher content of pectic substances and lower total dietary fibres as compared to "gading". The unique aroma of salak is the result of a complex composition of volatile compounds, some of them being still unidentified. The major aroma compounds of salak fruits consist of methyl esters of C5 and C6 branched-chain alkanoic, alkenoic and hydroxy alkanoic acids with their isomers. These patterns seem to be unique as compared to other tropical fruit species.

"Pondoh manggala" is the most interesting cultivar for the production due to its outstanding consumer acceptance. "Gading" seems not to be appropriate for the fresh fruit market. However, the high content of nutritional valuable dietary fibres in "gading" should be considered for the supply for the local population of the production areas.

In another study on the postharvest quality of salak fruits, physical and physiological changes during fruit maturation and ripening (4 until 6 months after pollination) of two salak cultivars, "pondoh hitam" and "pondoh super" were investigated. This was conducted in order to determine the quality related dynamics in salak fruit during postharvest. These informations will be the basis for optimising postharvest handling and storage operations and extending the marketing period with special emphasis on a consumer-oriented fruit quality.

Different patterns of peel and pulp colour changes were found in "pondoh super" and "pondoh hitam" during ripening. Physiological processes in "pondoh super" occurred to a later stage but then accelerated faster than "pondoh hitam" in terms of changes of mono- and disaccharides, resulting in a poorer marketability and shelf life. In respect to the sugar/acid ratio, a faster ripening process in "pondoh super" than in "pondoh hitam" was observed. "Pondoh super" had higher contents of polysaccharides and lignin, however the ripening processes occurred earlier in comparison to "pondoh hitam". The alterations in cell wall and middle lamella structure were not associated with the physical texture measurement during ripening.

Fruits of salak "pondoh hitam" and "pondoh super" can be harvested from 4 to 6 months after pollination, depending on the market orientation. Thus, if the target is to achieve optimal sensory attributes and nutritional contents, harvest time could be delayed in order to improve those quality compounds. Fruits at ripening stage 6 reached maximum size and contents of flavour components, but have lower mineral concentrations and structural carbohydrates in comparison to fruits at an earlier ripening stage. On the other hand, if minerals and structural carbohydrates are the predominant criterion for the market, i.e. for processing purposes, salak fruits should be harvested at an earlier ripening stage (stage 4).

Considering the results of these studies, future research activities should be directed towards the following aspects:

- Ecophysiological field studies with productive salak plants evaluating the interactive effects of soil type and nutrient availability, or fertiliser application and irrigation. The field studies should include the determination of fruit quality parameters

- Evaluation the ecophysiology and postharvest quality of other promising cultivars, such as "gula pasir" from Bali, in order to extend the genetical base for high quality salak fruit production.

- Further aroma compound studies in salak fruits with regard to some unidentified compounds. In the future, emphasis should also be placed on the reduction of the amount of undesired compounds such as 2-methylbutanoic acid and presumably other carboxylic acids in order to improve the acceptability of salak fruit worlwide.

- To evaluate simple, fast and non destructive methods for the determination of fruit quality, i.e. fruit firmness, which is the most important indicator for the optimal ripening and harvest stage of the fruit, which is important for the farmer.

8. SUMMARY

Salak is regarded to be very important for the Indonesian fresh fruit market. It also has a high potential for the export market. Among the salak cultivars, "pondoh" is an interesting one due to its superior quality. As a result of the major problems of salak production in Indonesia, e.g. fruit quality, quantity of yield and continuity of supply, an interdisciplinary approach in improving salak production techniques and postharvest properties is required. The objectives of these studies were to investigate the ecophysiological requirements of different salak cultivars, to investigate postharvest quality aspects of salak fruits and to assist the establishment of a quality-oriented production and postharvest management of salak fruit. A comprehensive review on salak plant and fruit including current investigation on production techniques and postharvest properties was conducted. The results of the study on the ecophysiology aspects as well as postharvest quality of different salak cultivars, "gading", "pondoh super", "pondoh hitam" and "pondoh manggala" can be summarised as follows:

- "Pondoh hitam" is the most vigorous cultivar tested on different growing media, such as sand, peat and sand/compost mixture. In general, the best growing medium for salak seedlings growth is sand, supplied with a complete nutrition solution. Peat with a very low pH was not suitable as growing medium for salak seedlings. Net CO_2 assimilation rate (P_N) of salak seedlings was relatively low as compared with other tropical tree species. Among the cultivar tested, "gading"exhibited the highest P_N. Foliar mineral contents (N, K, Ca, Mg) of salak seedlings were within the optimal range for growth in all media tested.

- Shading, which reduced sun light by 70 %, was not beneficial for 7-months-old salak seedlings. Shading conditions to certain extent are needed for raising salak seedlings due to photoinhibitory responses of non-shaded plants, which caused bleaching of salak leaves. Salak plants did not tolerate drought conditions. However, different levels of water supply did not affect P_N and mineral uptake (N, P, K, Ca,

Mg) of salak plants. Light had a stronger effect on P_N and nutrient uptake of salak seedlings than water supply.

- Salak fruits posses a high nutritional value with regard to soluble solid contents, disaccharides and minerals (K, Ca, Mg), but are relatively low in organic acids. With regard to contents of glucose and fructose, salak fruits are comparable to other tropical fruits. In respect to total dietary fibre, salak fruits are similar to temperate-zone fruits. Genetical factors (i.e. cultivar) have a strong effect on the nutritional value of salak fruit. Since "gading" is not closely related to the "pondoh" cultivars, the two groups of cultivars had significant differences in their nutrient value compounds. "Pondoh" fruits were characterised by sweeter taste, higher content of pectic substances and less total dietary fibres content in comparison to "gading". The unique aroma of salak is the result of a complex composition of volatile compounds, some of them being still unidentified. The major aroma compounds of salak fruits consist of the methyl esters of C5 and C6 branched-chain alkanoic, alkenoic and hydroxy alkanoic acids with their isomers. These patterns seem to be unique as compared to other tropical fruit species. "Pondoh manggala" is the most interesting cultivar for the production due to its outstanding consumer acceptance. "Gading" seems not to be appropriate for the fresh fruit market.

- Different physiological patterns between "pondoh super" and "pondoh hitam" were found with respect to changes of mono- and disaccharides, sugar/acid ratio and polysaccharides, resulting in a poorer marketability and shorter shelf life of "pondoh super" compared to "pondoh hitam". Alterations in the structure of the cell wall and middle lamella were not associated with the physical textural changes during ripening. The fruits of salak "pondoh hitam" and "pondoh super" cultivars can be harvested 4 to 6 months after pollination depending on the market orientation.

9. ZUSAMMENFASSUNG

Salak (*Salacca zalacca* (Gaertn.) Voss) ist eine wichtige Obstart auf dem indonesischen Markt und besitzt ein hohes Exportpotenzial. Um bestehende Probleme der Salakproduktion, wie schlechte Fruchtqualität, geringe Erträge und ungleichmäßiges Angebot im Jahresverlauf zu verringern, ist ein interdisziplinärer Lösungsansatz notwendig. In diesem Kontext hat die vorliegende Arbeit das Ziel, zur Verbesserung der Anbaubedingungen von Salakpalmen in Indonesien sowie zur Erhöhung der Fruchtqualität beizutragen. Dazu wird zunächst ein Überblick über botanische Besonderheiten der Obstart Salak gegeben und die Produktion in Indonesien einschließlich der Nacherntephase vorgestellt. Anschließend werden in Berlin durchgeführte Untersuchungen zu ökophysiologischen Aspekten von Salaksämlingen beschrieben. Einen weiteren Teil dieser Arbeit stellen Untersuchungen zu Fruchtinhaltsstoffen und -eigenschaften von ausgewählten Salaksorten („Gading", „Pondoh Super", „Pondoh Hitam" und „Pondoh Manggala") dar, die in Berlin und Quedlinburg durchgeführt wurden.

Das optimale Wurzelsubstrat für die Topfkultur war Quarzsand, wenn die Pflanzen mit einer Vollnährlösung versorgt wurden. Für die Anzucht von Jungpflanzen ist bei starker Sonneneinstrahlung eine Schattierung (Einstrahlung bis ca. 700 μMol $m^{-2}s^{-1}$) notwendig, da die Blätter sonst schnell ausbleichen. Zu starke Schattierung (weniger als 250 μMol m^{-2} s^{-1}) führte jedoch zu geringen Wuchsleistungen der Pflanzen. Salaksämlinge tolerierten keine Trockenphasen während der Anzucht. Die Photosyntheserate von Salaksämlingen ist im Vergleich zu anderen tropischen Obstarten als gering einzustufen.

Untersuchungen an frisch eingeführten Früchten ergaben, dass diese im Vergleich zu anderen Obstarten einen sehr hohen Zuckergehalt besitzen und relativ viele Mineralstoffe (K, Ca, Mg) anreichern. Sie sind säurearm, während sich die weiteren Fruchtinhaltsstoffe kaum von denen anderer tropischer Obstarten unterscheiden. Ein Testpanel beurteilte die sensorischen Eigenschaften der Früchte. „Pondoh"-Sorten waren süßer und hatten höhere Gehalte an Pektin als „Gading". Von den untersuchten

Sorten wurde „Pondoh Manggala“ als beste eingestuft, wohingegen „Gading“ wegen fehlender Süße und unangenehmer Geruchs- und Geschmackseigenschaften als Frischfrucht und für den Export nicht in Frage kommt. Das charakteristische „Salak-Aromastoffe“ der Früchte wird im Wesentlichen von Methylestern einiger Alkansäuren, Alkensäuren und Hydroxialkansäuren sowie ihrer Isomere gebildet.

Es wurden sortenspezifisch unterschiedliche Reifeverläufe bezüglich der Entwicklung von Mono- und Disacchariden, Polysacchariden und dem Zucker/Säure-Verhältnis gefunden, die auf ein kürzeres shelf-life von „Pondoh Super“ im Vergleich zu „Pondoh Hitam“ hinweisen. Veränderung der Zellwand- und Mittellamelle waren nicht mit physikalischen Veränderung der Fruchtfleischtextur verbunden. Je nach Verwendung sollten Früchte der Sorten „Pondoh Hitam“ und „Pondoh Super“ 4 bis 6 Monate nach dem Fruchtansatz geerntet werden.

10. REFERENCES

Acree, T. and Arn, H. 1997. Flavornet. Cornell University. USA http://www.nysaes.cornell.edu/fst/faculty/acree/flavornet/chem.html

Agren, G.I. 1985. Theory for growth of plants derived from the nitrogen productivity concept. Physiol. Plant. 64, 17-28

Agren, G.I. and Ingestand, T. 1987. Root:shoot ratio as a balance between nitrogen productivity and photosynthesis. Plant Cell Environ. 10, 579-586

Amiarsi, D., Sitorus, E. and Sjaifullah 1996. The effect of temperature on the quality of *Salacca* during storage (in Indonesian). Laporan Penelitian. Badan Penelitian dan Pengembangan Pertanian. Departemen Pertanian. Jakarta, Indonesia

Andriesse 1988. Nature and management of tropical peat soils. FAO Soils Bulletin 59. Food and Agriculture Organisation of The United Nations, Rome, Italy

Anonymous 2002. Flavors and fragrances: The essence of excellence. Sigma Aldrich, Fine Chemicals, International Edition 2001-2002

AOAC 1984. Official methods of analysis (14th Edn.), Association of Official Analytical Chemists, Washington DC, USA

AOAC 1990. Official methods of analysis (15th Edn.), Association of Official Analytical Chemists, Arlington, VA, USA

Arintadisastra, S. 1997. The policy and constraint on the development of fruit production (in Indonesian). Paper presented at the workshop on the development prospect of fruit agribusiness, 5 February. Bogor, Indonesia

Ashari 1995. The storage of salak pollen (*Salacca zalacca* (Gaertn.) Voss) (in Indonesian). J. Univ. Brawijaya 7 (3), 40-44

Asher, C.J. 1978. Natural and synthetic culture media for Spermatophytes. CRC Handb. Ser. Nutr. Food, Sect. G 3, 575-609

Axley, J.H. and Legg, J.O. 1960. Ammonium fixation in soils and the influence of potassium on nitrogen availability from nitrate and ammonium sources. Soil. Sci. 90, 151-156

Backer, C.A. and van den Brink, Jr. R.C.B. 1968. Flora of Java. Vol. III. Groningen. The Netherlands

Bailey, L.H. 1946. The palm herbarium. Gentes. Herb. 7, 178

Baswarsiati, L., Roesmahani and Setyobudi, L. 1993. Study on the insects in salak pollination (in Indonesian). Penel. Hort. 5 (2), 45-71

Baswarsiati, L. and Rosmahani, L. 1994. The potency of *Curculionidae* as pollinators for the snake fruit (in Indonesian). J. Hort. 4 (2), 55-58

Batten, D.J. 1989. Maturity criteria for litchis (lychees). Food Qual. Prefer. 1, 149-155

Berliana, Y. 2002. Optimalising of sucrose concentration and plantlet pre-adaptation of salak (*Salacca sumatrana* Becc.) in low humidity (in Indonesian). J. Penel. Pert. 21 (1), 64 –73

Biro Pusat Statistik. 1995. Fruit production in Indonesia (in Indonesian). BPS. Jakarta, Indonesia

Biswal, B. and Biswal, U.C. 1999. Photosynthesis under stress: Stress signals and adaptive response of chloroplast. *In*: Pessarakli, M. (ed.). Handbook of plant and crop stress (2nd ed., revised and expanded). Marcel Dekker, Inc. New York and Basel, 315 – 336

Björkman, O. and Powles, S.B. 1984. Inhibition of photosynthesic reactions under water stress: Interaction with light level. Planta 161, 490-504

Blumenkrantz, N. and Asboe-Hansen, G. 1973. New method for quantitative determination of uronic acids. Anal. Biochem. 54, 484-489

Bollard, E.G. 1970. Physiology and nutrition of developing fruit. *In*: Hulme, A.C. (ed.). The biochemistry of fruits and their products. Academic Press London and New York, 387-424.

Bowo, H. 1999. Gamet fertilisation in salak plant (*Salacca zalacca* (Gaertn.) Voss) (in Indonesian). MIP UPN Veteran Jawa Timur 8 (20), 105-110

Boyer, J.S. 1976. Water deficit and photosynthesis. *In*: Kozlowski, T.T. (ed.). Water deficits and plant growth, Vol. IV: Soil water mesurements, plant responses and breeding for drought resistance, Academic Press, London and New York, 153-190

Brakke, M. and Allan Jr, L.H. 1995. Gas exchange of Citrus seedlings at different temperatures, vapour-pressure deficits and soil water contents. J. Am. Soc. Hort. Sci. 120, 497-504

Buwalda, J.G. and Meekings, J.S. 1990. Seasonal accumulation of mineral nutrients in leaves and fruit of Japanese pear (*Pyrus serotina* Rehd.), Sci. Hort. 41 (3), 209-222

Cattivello, C., Della Donna, E. and Pantanali, R. 1997. Behaviour of peat substrates during cyclamen and poinsettia cultivation. Acta Hort. 450, 439-447

Chaney, R.L., Munns, J.B. and Cathey, H.M. 1980. Effectiveness of digested sewage sludge compost in supplying nutrients for soilless potting media. J. Am. Soc. Hort. Sci. 105 (4), 485-492

Cho, S., De Vries, J. W. and Prosky, L. 1997. Dietary fiber analysis and applications, Association of Official Analytical Chemists International. Maryland, USA

Coombe, B.G. 1962. The effect of removing leaves, flowers and shoot tips on fruit-set in *Vitis vinifera* L. J. Hort. Sci. 37, 1-15

Corner, E.J.H. 1966. The Natural History of Palms. Berkeley. USA

Cornic, G. 2000. Drought stress inhibits photosynthesis by decreasing stomatal aperture not by affecting ATP synthesis. Trends Plant Sci. 5, 187-188

Crookston, R.K., Treharne, K.J., Ludford, P. and Ozbun, J.L. 1975. Response of beans to shading. Crop. Sci. 15, 412-416

Dangcham, S. and Siripanich, J. 2001. Quality variation and storage of sala fruits cv. *Nerwong*. Acta Hort. 2 (553), 741-743

DeJong, T.M. and Doyle, J.F. 1985. Seasonal relationships between leaf nitrogen content (photosynthetic capacity) and leaf canopy light exposure in peach (*Prunus persica*). Plant Cell Environ. 8, 701-706

Dendi, H. 1997. Export chances of horticultural products (in Indonesian). Panitia Dies Natalis ke–51. Fakultas Pertanian. Universitas Gadjah Mada. Yogyakarta, Indonesia

Departemen Pertanian. 1998. Distribution and marketing of salak pondoh in Yogyakarta province (in Indonesian). Departemen Pertanian. Yogyakarta, Indonesia

Departemen Pertanian. 2001. Statistics of fruit plants in the Sleman district (in Indonesian). Laporan Tahunan 1995-2001. Departemen Pertanian. Yogyakarta, Indonesia

Dickie, J.B., Balick, M.J. and Linington, I.M. 1993. Studies on the practicability of *ex situ* preservation of palm seeds. Principes 37 (2), 94-98

Dickson, R.E. and Isebrands, J.G. 1991. Leaves as regulators of stress responses. *In*: Money, H.A., Winner, W.E. and Pell, E.J. (eds.). Response of plants to multiple stresses. Academic Press, Inc. California, USA, 3-34

Djaafar, T.F. and Thamrin, M. 1996. The inventory of some salak pondoh cultivars in Sleman district, Yogyakarta province (in Indonesian). Paper presented at "Apresiasi Program Pembangunan Pertanian Tingkat Propinsi D.I. Yogyakarta", 19-20 November. Yogyakarta, Indonesia

Djaafar, T.F. 1998. Salak pondoh, superior commodity from Yogyakarta province and the postharvest handling (in Indonesian). Paper presented at "Jumpa Teknologi Hortikultura di Dinas Pertanian Tanaman Pangan Propinsi D.I.Y., 2 December. Yogyakarta, Indonesia

Djaafar, T.F. and Mudjisihono, R. 1998. Study on the optimal harvesting date of salak pondoh fruit to support the salak cultivation systems in Yogyakarta province (in Indonesian). Prosiding Seminar Ilmiah dan Lokakarya, Departemen Pertanian. Yogyakarta, Indonesia, 210 - 216

Dransfield, J. and Uhl, N.W. 1986. An outline of a classification of Palms. Principes 30, 3-11

Dufrene, E. and Saugier, B. 1993. Gas exchange of oil palm in relation to light, vapour pressure deficit, temperature and leaf age. Functional Ecol. 7, 97-104

Dunlop, J. and Bowling, D.J.F. 1978. Uptake of phosphate by white clover. II. The effect of pH on the electronic phosphate pump. J. Exp. Bot. 29, 1147 – 1153

Ellis, C. and Swaney, M.W. 1938. Soilless growth of plants. Reinhold Publishing Corporation. New York, USA

Evenhuis, B. and de Waard, P.W.F. 1980. Principles and practices in plant analysis. FAO Soils Bulletin 38 (1), 152-163

Fairhorst, T. and Härdter, R. 2003. Oil palm – Management for large and sustainable yields. Potash and Phosphate Institute (PPI), Potash and Phosphate Institute of Canada (PPIC) and International Potash Institute (IPI). Singapore

Fisher J.B. and Mogea, J.P. 1980. Intrapetiolar inflorescence buds in *Salacca* (*Palmae*): Development and significance. Bot. J. Linnean Soc. 81, 47-59

Fitzpatrick, G.E. 2002. Compost utilization in ornamental and nursery crop production systems. *In*: Stofella, P.J. and Kahn, B.A. (eds). Compost utilization in horticultural cropping systems. Lewis Publishers, CRC Press LLC, Florida, USA

Fredeen, A.L., Rao, I.M. and Terry, N. 1989. Influence of phosphorous nutrition on growth and carbon partitioning in *Glycine max*. Plant. Physiol. 89, 225 - 230

Fukamachi, H., Yamada, M., Komori, S. And Hidaka, T. 1998. Photosynthesis in longan and mango as influenced by high temperatures under high irradiance. JIRCAS Newsletter No. 17

Furtado, C.K. 1949. Palmae malesiae-X. The Malayan species of *Salacca*. Gard. Bull. Straits Settlem. 12, 378-403

Giovannoni, J.J., DellaPenna, D., Bennet, A.B. and Fisher, L.R. 1989. Expression of a chimeric polygalacturonase gene in transgenic *rin* (ripening-inhibitor) tomato fruit results in polyuronide degradation but not fruit softening. Plant Cell 1, 53-56

Goering, H.K. and van Soest, P.J. 1972. Forage fiber analyses (apparatus, reagents, procedures and some applications), Agriculture Handbook 379. Washington, USA

Grime, J.P. and Campbell, B.D. 1991. Growth rate, habitat productivity, and plant strategy as predictors of stress response. *In*: Mooney, H.A., Winner, W.E. and Pell, E.J. (eds.). Response of plant to multiple stresses. Academic Press. London, UK

Gross, K.C. and Walner, S.J. 1979. Degradation of cell wall polysaccharides during tomato fruit ripening, Plant Physiol. 63, 117-120

Hadi, P.S. 2001. Chromosome identification for sex determination in *Salacca zalacca* (Gaertn.) Voss (in Indonesian). Master Thesis. Program Pasca Sarjana Universitas Gadjah Mada. Yogyakarta, Indonesia

Handreck, K.A. and Black, N.D. 1999. Growing media for ornamental plants and turf. UNSW Press Book. Sydney. Australia

Harman, J.E. 1981. Kiwifruit maturity. Orchard. N. Z. 54, 126 - 130

Hartanto, R., Rahardjo, B. and Suhardi 2000. Modelling the di- and monosaccharide change of salacca (*Salacca edulis* Reinw. cv *pondoh*) fruit at modified atmosphere condition (in Indonesian). Agritech 20 (1), 10-13

Hastuti, P. and Ari, M. 1988. Changes in the chemical and consumer preference of salak pondoh during cold storage (in Indonesian). Prosiding Seminar Penelitian Pascapanen Pertanian. Badan Penelitian dan Pengembangan Pertanian, Departemen Pertanian. Bogor, Indonesia

Hayasaka, Y., MacNamara, K., Baldock, G. A., Taylor, R. L. and Pollnitz, A. P. 2003. Application of stir bar sorptive extraction for wine analysis. Anal. Bioanal. Chem. 375, 948-955

Herppich, M., Herppich, W.B. and von Willert, D.J. 1994. Influence of drought, rain and artificial irrigation on photosynthesis, gas exchange and water relations of the fynbos plant *Protea acaulos* (L.) Reich at the end of the dry season. Bot. Acta 107, 369-472

Herppich, W.B. 2000. Interactive effects of light and drought stress on photosynthesic activity and photoinhibition under (sub-) tropical conditions. Acta Hort. 531, 135-142

Herrmann, K. 2001. Inhaltsstoffe von Obst und Gemüse, Verlag Eugen Ulmer, Stuttgart, Germany

Hewitt, E.J. 1952. Sand and water culture methods used in the study of plant nutrition. Commonwealth Agricultural Bureaux. UK

Hirose, T. 1988. Modelling the relative growth rate as a function of plant nitrogen concentration. Physiol. Plant. 72, 185-189

Hirose, T. and Werger, J.A. 1987. Nitrogen use efficiency in instantaneous and daily photosynthesis of leaves in the canopy of a *Solidago altissima* stand. Physiol. Plant. 70, 215-222

Hirose, T., Werger, J.A., Pons, T.L. and van Rheenen, J.W.A. 1988. Canopy structure and leaf nitrogen distribution in a stand of *Lysimachia vulgaris* L. as influenced by stand density. Oecologia 77, 145-150

Hobson, G. and Grierson, D. 1993. Tomato. *In*: Seymour, G.B., Taylor, J.E. and Tucker, J.E. (eds.). Biochemistry of Fruit Ripening, Chapman & Hall, London, UK, 405-442

Hsiao, T.C. 1973. Plant responses to water stress. Ann. Rev. Plant Physiol. 24, 519 - 570

Hsiao T.C., Acevedo, E., Fereres, E. and Henderson, D.W. 1976. Water stress, growth and osmotic adjustment. Philos. Trans. R. Soc. Lond. Ser. B, 273, 479-500

Huber, D.J. 1983. Polyuronide degradation and hemicellulose modification in ripening tomato fruit. J. Am. Soc. Hort. Sci. 108, 405-409

Hutauruk, H.D. 1999. The development of salak bali seed (*Salacca zalacca* var *amboinensis*) (in Indonesian). Master Thesis. Program Pasca Sarjana, Institut Pertanian Bogor. Bogor. Indonesia

Huyskens, S. 1991. Morphological, physiological, and biochemical aspects in the cultivation of two pantropical cucurbits: *Luffa acutangula* (L.) Roxb. and *Momordica charantia* L. Dissertation, Universität Bonn, Germany

Huyskens-Keil, S. and Schreiner, M. 2003. Quality dynamics and quality assurance of fresh fruit and vegetable products in pre- and postharvest. *In*: Dris, R., Niskanen, R., and Jain, S.M. (eds.), Production practices and quality assessment of food crops. Vol. 3: Harvest and quality evaluation. Science Publishers, Inc., Enfield, USA

Imad, A.A., Abdul Wahab, K.A. and Robinson, R.K. 1995. Chemical composition of date varieties as influenced by the stage of ripening. Food Chem. 54, 305-309

Ina, P.T. 1997. Postharvest technology and processing of salak fruit (in Indonesian). Risalah Hasil Penelitian Peternakan dan Tanaman Pangan. Departemen Pertanian. Jakarta, Indonesia

Ismail, M.R., Burrage, S.W., Tarmizi, H. and Aziz, M.A. 1994. Growth, plant water relation, photosynthesis rate and accumulation of proline in young carambola plants in relation to water stress. Sci. Hort. 60, 101-114

Jackman, R.L. and Stanley, D.W. 1995. Review: Perspective in the textural evaluation of plant foods. Trends Food Sci. Technol. 6, 187-194

Jellinek, G. 1985. Sensory evaluation of food, theory and practice. Ellis Horwood Ltd., Chichester, England

Jezussek, M., Juliano, B. O. and Schierberle, P. 2002. Comparison of key aroma compounds in cooked brown rice varieties based on aroma extract dilution analyses. J. Agric. Food Chem. 50, 1101-1105

Jirovetz, L., Smith, D. and Buchbauer, G. 2002. Aroma compound analysis of *Eruca sativa* (Brassicaceae) SPME headspace leaf samples using GC, GC-MS, and olfactometry. J. Agric. Food Chem. 50, 4643-4646

Jones, H.G. 1992. Plants and microclimate, a quantitive approach to environmental plant physiology (2nd ed.). Cambridge University Press. UK

Jones, H.G. 1998. Stomatal control of photosynthesis and transpiration. J. Ex. Bot., Special Issue 49, 387-398

Jordan, M. J., Margaria, C. A., Shaw, P. E. and Goodner, K. L. 2003. Volatile components and aroma active compounds in aqueous essence and fresh pink guava fruit puree (*Psidium guajava* L.) by GC-MS and multidimensional GC/GC-O. J. Agric. Food Chem. 51, 1421-1426

Kader, A.A. and Barrett, D.M. 1996. Classification, composition of fruits and postharvest maintenance of quality. *In*: Somogyi, L.P., Ramaswamy, H.S. and Hui, Y.H. (eds.), Processing fruits: Science and technology. Vol. 1: Biology, principles and applications. Technomic Publishing Co. Inc. Lancaster Basel, Switzerland, 1-24

Kasijadi, F., Purbiati, T., Mahfud, M.C., Sudaryono, T. and Soemarsono, S.R. 1999. The application air-layering propagation *Salacca* seedlings (in Indonesian). J. Hort. 9 (1), 1–7

Kusmiba, P. 1994. The effect of growing media and foliar fertilisation to growth of salak seedlings in peat land (in Indonesian). J. Penel. UNTAN 4 (14), 25-33

Kusumainderawati, E.P. and Soleh, M. 1995. Standard determination of fertiliser demand for growth and production of salak (in Indonesian), J. Hort. 5 (2), 23-29

Kusumo, S. 1995. Salak, a prideful fruit of Indonesia. IARD Journal 17 (2), 19–23

Le Bot, J., Adamowicz, S. and Robin, P. 1998. Modelling plant nutrition of horticultural crops: A review. Sci. Hort. 74, 47-82

Leong, L.P. and Shui, G. 2002. An investigation of antioxidant capacity of fruits in Singapore markets. Food Chem. 76, 69-75

Lestari, R. and Ebert, G. 2002. Salak (*Salacca zalacca* (Gaertn.) Voss) – The snakefruit from Indonesia, preliminary results of an ecophysiological study. Proceeding of Deutscher Tropentag 2002, Kassel, Germany

Lestari, R. and Ebert, G. 2003. The snake fruit salak pondoh (*Salacca zalacca* (Gaertner) Voss.) – A new fruit species from Indonesia. BDGL-Schriftenreihe 21: 208

Lestari, R., Huyskens-Keil, S. and Ebert, G. 2003. Quality aspects of salak pondoh fruit (*Salacca zalacca* (Gaertn.) Voss) from Indonesia. BDGL-Schriftenreihe 21:120

Lestari, R., Huyskens-Keil, S. and Ebert, G. 2004. Physiological changes of salak fruit (*Salacca zalacca* (Gaertn.) Voss) during maturation and ripening. BDGL-Schriftenreihe 22: 201

Lestario, L.N., Adam, N., Wulansari, R.C. and Pratiani, F.E. 1999. Modified atmosphere packaging of apple manalagi and salak pondoh (in Indonesian). Prosiding Seminar Nasional Teknologi Pangan, Jakarta, 12-13 Oktober, 624-631

Lestyana, B. 2000. Multiplication of salak cv gula pasir (*Salacca zalacca*) by *in vitro* and float sub culture method (in Indonesian). Master Thesis. Program Pasca Sarjana. Institut Pertanian Bogor. Bogor, Indonesia

Leudauphin, J., Guichard, H., Saint-clair, J. F., Picoche, B. and Barillier, D. 2003. Chemical and sensorial aroma characterization of freshly distilled calvados. 2. Identification of volatile compounds and key odorants. J. Agric. Food Chem. 51, 433-442

Levitt, J. 1980. Responses of plants to environmental stress. 2nd ed. Academic Press, New York, USA

Lim, J.T., Wilkerson, G.G., Raper, C.D. Jr. and Gold, H.J. 1990. A dynamic growth model of vegetative soya bean plants: Structure and behaviour under varying root temperature and nitrogen concentration. J. Exp. Bot. 41, 229-241

Lodh, S.B. Divakar, N.G., Chada, K.L. and Melanta, K.R. 1972. Biochemical changes associated with growth and development of pineapple fruit variety Kew II. Changes in carbohydrate and mineral constituents. Indian J. Hort. 29, 287-291

Lüttge, U. 1997. Physiological ecology of tropical plants. Springer. Berlin-Heidelberg, Germany

MacLeod, A. J. and deTroconis, N. G. 1982. Volatile flavor components of cashew apple (*Anacardium occidentale*). Phytochem. 21, 2527-2530

Mahendra, M.S. and Janes, J. 1993. Indice of chilling injury in *Salacca zalacca*, *In*: Champ, B.R., Highley, E. and Johnson, G.I. (eds.). Postharvest handling of tropical fruits. Proceedings of an International Conference held at Chiang Mai, Thailand, 19–23 July, 402-404

Mahfud, M.C., Roesmahani, L. and Sidik, N.I. 1993. Identification and protection potency of salak pest and disease (in Indonesian). Penel. Hort. 5 (2), 56-71

Mahfud, M.C., Roesmahani, L. and Sidik, N.I. 1994. Bionomy of *silphidae* and leaf spot disease (*Pestaliopsis palmarum*) in salak (in Indonesian). Penel. Hort. 6 (2), 29-39

Maier, H.G. 1990. Lebensmittel- und Umweltanalytik. Steinkopff Verlag, Darmstadt, Germany

Mangoendihardjo, S. 1975. *Curculionidae* and its protection from salak plants in Sleman district of Yogyakarta province (in Indonesian). Prosiding Seminar Biologi IV, Yogyakarta, Indonesia

Manrique, G.D. and Lajolo, F.M. 2002. Cell-wall polysaccharides modifications during postharvest ripening of papaya fruit (*Carica papaya*), Postharvest Biol. Technol. 33, 11-26

Marschner, H. 1995. Mineral nutrition of higher plants. 2nd. ed.. Academic Press. CA, USA

McComb, E.Y. and McCready, R. 1952. Colorimetric determination of pectic substances. Anal Chem. 24 (10), 1630-1632

McGuire, R. 1992. Reporting of objective color measurements. Hort. Sci. 27, 1254-1255

Miller, R.H. 1978. Fruit, germination and developmental morphology of *Salacca edulis* palm seedling. Phytomorphology 27 (3), 282-296

Mogea, J.P. 1978. Pollination in *Salacca edulis*. Principes, 22 (2), 56–63

Mogea, J.P. 1979. Fruit production of the salak palm and its relationship to average annual rainfall (in Indonesian). Berita Biol. 2 (4), 71–74

Mogea, J.P. 1982. *Salacca zalacca*, the correct name for the salak palm. Principes 26 (2), 70-72

Mogea, J.P. 1992. Progress on the taxonomy of the genus *Salacca* and *Arenga*. Presented paper in the Second Intern. Symp. of the Flora Malesiana, 7–12 September, Yogyakarta, Indonesia

Moncur, M.W. and Watson, B.J. 1987. Observation on the floral biology of the monoecious form of *Salacca zalacca*. Principes, 31 (1), 20 – 22

Mudjisihono, R. and Handayani 2000. The effect of additional carboxy methyl cellulose (CMC) on the physical and sensory properties of salak juice during storage (in Indonesian). J. Penel. Pert. 19 (1), 56–67

Murayama, H., Katsumata, T., Horiuchi, O. and Fukushima, T. 2002. Relationship between fruit softening and cell wall polysaccharides in pears after different storage periods, Postharvest Biol. Technol. 26, 15-21

Murtiningsih, W., Prabawati, S. and Sjaifullah 1995. Pathogen-caused postharvest disease in *Salacca* fruit and its control (in Indonesian). Laporan Hasil Penelitian. Pusat Penelitian dan Pengembangan Hortikultura. Departemen Pertanian. Jakarta, Indonesia

Ochse, J.J. 1931. Fruits and fruitculture in the Dutch East Indies (English eds.). G. Kolff & Co – Batavia, Indonesia

Ochse, J.J., Soule, M.J., Dijkman, M.J. and Wehlburg, C. 1961. Tropical and subtropical agriculture. Macmillan, New York, USA

Okoye, H.C. 1980. Plant analysis as an aid in the fertilization of the oil palm. FAO Soils Bulletin 38 (1), 164-179

Oliviera, M.A.J., Bovi, M.L.A., Machado, E.C., Gomes, M.M.A., Habermann, G. and Rodrigues, J.D. 2002. Photosynthesis, stomatal conductance and transpiration in peach palm under water stress. Sci. Agric. (Piracicaba Braz.), 59 (in Press)

Padmosudarso, S. 2000. Land suitability for salak pondoh in subdistrict Turi, sleman district, Yogyakarta province (in Indonesian), Disertasi Program Doktor, Fakultas Pasca Sarjana, Universitas Gadjah Mada, Yogyakarta, Indonesia

Page, P.E., 1984. Tropical tree fruits for Australia. Queensland Department of Primary Industries. Brisbane, Australia

Partha, I.B.B., Tranggono and Haryadi. 1993. The properties of salak pectins at various storage temperatures (in Indonesian). BPPS-UGM, 6 (3B), 339-351

Paull, R.E. 1982. Postharvest variation in composition of soursop (*Annona muricata* L.) fruit in relation to respiration and ethylene production. J. Am. Soc. Hort. Sci. 107, 582-585

Pinamonti, F. and Sicher, L. 2000. Compost utilization in fruit production systems. *In*: Stoffella, P.J. and Kahn, B.A. (eds.). Compost utilization in horticultural cropping systems. Lewis Publishers, CRC Press LLC, Florida, USA

Pollien, P. Ott, A., Montigon, F., Baumgartner, M., Munoz-Box, R. and Chaintreau, A. 1997. Hyphenated-gas chromatography-sniffing technique: Screening of impact odorants and quantitative aromagram comparison. J.Agric. Food Chem. 45, 2630-2637

Polprasid, P. 1992. *Salacca walichiana* C. Martius. *In*: Verheij, E.W.M. and Coronel, R.E. (eds.). Edible fruits and nuts. Plant Resources of South East Asia No. 2. PROSEA. Bogor, Indonesia

Powles, S.B. 1984. Photoinhibition of photosynthesis induced by visible light. Ann. Rev. Plant Physiol. 35, 15-44

Prabawati, S., Utami, A.D. and Sjaifullah 1994. The effect of ascorbic-bisulfit and benzoate treatments on quality and damage of peeled salak pondoh (in Indonesian). Penel. Hort. 6 (2), 74–84

Prabawati, S., Sujanti and Sjaifullah 1996. Determination of the proper maturity of Salacca fruits cv suwaru to acquire good quality of fruits (in Indonesian). J. Hort. 6 (3), 309-317

Prahardini, P.E.R., Sudaryono, T and Purnomo, S. 1991. The effect of media composition and explant source to initiation and proliferation of *in vitro* cultured salak (in Indonesian). Laporan Penelitian. Sub Balai Penelitian Hortikultura Malang. Departemen Pertanian. Jawa Timur, Indonesia

Prahardini, P.E.R., Sudaryono, T. and Tegopati, B. 1993. The effect of cytokinin application to MS media on the multiplication of shoots (in Indonesian). Laporan Hasil Penelitian. Sub Balai Penelitian Hortikultura Malang. Departemen Pertanian. Jawa Timur, Indonesia

Prahardini, P.E.R. and Sudaryanto, T. 1995. *In vitro* propagation of salak (in Indonesian). Warta Penel. dan Pengemb. Pert. 17 (4), 8-10

Pugnaire, F.I., Serrano, L. and Pardos, J. 1999. Constraints by water stress on plant growth. *In*: Pessarakli, M. (ed.). Handbook of plant and crop stress (2nd ed., revised and expanded). Marcel Dekker, Inc. New York, and Basel, 271 – 283

Purbiati, T., Ernawanto, Q.D. and Soemarsono, S.R. 1993. Fast clonal propagation for the technological development of salak production and quality (in Indonesian). Laporan Penelitian Sub Balai Hortikultura Malang, Departemen Pertanian, Jawa Timur, Indonesia, 191-197

Purbiati, T., Ernawanto, Q.D. and Soemarsono, S.R. 1994. The effect of container size and media composition on air-layering propagation of salak (in Indonesian). Prosiding Seminar Hasil Penelitian Buah-buahan 1993/1994. Sub Balai Penelitian Hortikultura Malang. Departemen Pertanian. Jawa Timur, Indonesia, 179-188

Purbiati, T., Sumarsono, S.R. and Hermanto, C. 1999. Field test and financial analysis of *Salacca* nursery from Bali derived from air-layering and seed (in Indonesian). J. Hort. 9 (1), 59–66

Purnomo, S. and Sudaryono, T. 1994. Selection of high quality plant from the population of salak Bali and salak pondoh (in Indonesian). Laporan Hasil Penelitian. Sub Bali Penelitian Malang. Departemen Pertanian. Jawa Timur, Indonesia

Purnomo, S. and Dzanuri 1996. Heterosis analysis and production of hybrids seeds on crossing of salacca Bali and pondoh cultivars (in Indonesian). J. Hort. 6 (3), 233-241

Purseglove, J.W. 1968. Tropical Crops. Longman. London, UK

Purwadaria, H.K., Gunadya, I.B.P. and Fardiaz, D. 1992. Study on modified atmosphere packaging of salak (*Salacca edulis* Reinw.) (in Indonesian). Laporan Penelitian. Pusat Antar Universitas Pangan dan Gizi. Institut Pertanian Bogor. Bogor, Indonesia

Purwanto, Y., Rahayu, R.D. and Sutarno, H. 1988. The tolerance of salak seed to decrease of water content, temperature and fungi (in Indonesian). Berita Biol. 3 (8), 390-395

Rapp, A., Hastrich, H. and Engel, L. 1976. Gaschromatographische Untersuchung über die Aromastoffe von Weinbeeren. Vitis 15, 29-36

Reid, M.S. 2002. Maturation and maturity indices, *In*: Kader, A.A. (ed.). Postharvest technology of horticultural crops. University of California, Agriculture and Natural Resources, Publication 3311, USA

Rejo, A. 1996. Exchange of CO_2 and O_2 of salak pondoh in polyethylene packages during modified atmosphere storage (in Indonesian). Master Thesis. Program Pasca Sarjana. Universitas Gadjah Mada. Yogyakarta, Indonesia

Roosmahani, L. and Sjaifullah 1991. Storage of salak pondoh in a modified atmosphere system (in Indonesian). Sub Balai Penelitian Hortikultura Pasarminggu. Departemen Pertanian. Jakarta, Indonesia

Roosmahani, L. 1992. Packaging *Salacca* pondoh and Bali cultivars under modified atmosphere (in Indonesian). Laporan Hasil Penelitian. Badan Penelitian dan Pengembanagan Pertanian. Departemen Pertanian. Jakarta, Indonesia

Rosmahani, L., Baswarsiati and Sudaryono, T. 1993. Application and amount of effective population of pollinator insect in salak (in Indonesian). Laporan Penelitian. Sub Balai Penelitian Hortikultura Malang. Departemen Pertanian. Jawa Timur, Indonesia

Ross, M. 1999. Auswirkungen verschiedener Rodeverfahren und des Unterbewuchses auf Bodenfruchtbarkeit, Bodenwasserhaushalt, Erosion und Bestandsentwicklung eines Ölpalmenbestandes. Dissertation. Humboldt Universität zu Berlin. Berlin, Germany

Rukmana, R.H. 1999. Salak, agribussines and agricultural technique (in Indonesian). Kanisius. Yogyakarta, Indonesia

Sage, R.F. and Pearcy, R.W. 1987. The nitrogen use efficiency of C_3 and C_4 plants: II. Leaf nitrogen effects on the gas exchange characteristics of *Chenopodium album* (L.) and *Amaranthus retroflexus* (L.). Plant Physiol. 84, 954-958

Sakho, M., Chassagne, D. and Crouzet, J. 1997. African mango glycosidally bound volatile compounds. J.Agric. Food Chem. 46, 883-888

Salisbury, F.B. and Ross, C.W. 1985. Plant physiology. International Thompson Publishing, CA, USA

Santosa, T., Sujono and Rohadi, P.N., 1996a. Description of salak pondoh and hand pollination technique (in Indonesian). Departemen Pertanian. Yogyakarta, Indonesia

Santosa, T., Djaafar, T.F., Sukar and Aliudin 1996b. The inventory of some salak pondoh types in Sleman district, Yogyakarta province (in Indonesian). BPPT Ungaran, INPPTP D.I.Y. Yogyakarta, Indonesia

Santoso, B.H. 1990. Salak pondoh (in Indonesian). Kanisius. Yogyakarta, Indonesia

Schieberle, P. and Hofmann, T. 1997. Evaluation of the character impact odorant in fresh strawberry juice by quantitative measurements and sendory studies on model mixtures. J. Agric. Food Chem. 45, 227-232

Schimansky, C. 1981. Der Einfluss einiger Versuchparameter auf das Fluxverhalten von ^{28}Mg bei Gerstenkeimpflanzen in Hydrokulturversuchen. Landwirtsch. Forsch. 34, 154 -165

Schubert, S., Schubert, E. and Mengel, K. 1990. Effect of low pH of the root medium on proton release, growth, and nutrient uptake of field beans (*Vicia faba*). Plant Soil 124, 239-244

Schuiling, D.L. and Mogea, J.P. 1992. *Salacca zalacca* (Gaertner) Voss. In: Verheij, E.W.M. and Coronel, R.E. (eds.), Edible fruits and nuts, Plant Resources of South East Asia no. 2. PROSEA, Bogor, Indonesia, 247-248

Semarajaya, C.G.A. 1991. The effect of oxygen and carbondioxid on the storage life of Balinese salak cultivars (in Indonesian). Master Thesis. Program Pasca Sarjana. Universitas Gadjah Mada. Yogyakarta, Indonesia

Setiasih, I.S. 1999. Assessment of quality changes of edible coated minimally processed fruits (pondoh salak and arumanis mango) during storage (in Indonesian). Disertasi Program Doktor. Fakultas Pasca Sarjana. Institut Pertanian Bogor. Bogor, Indonesia

Setiawan, B., Sulaeman, A., Giraud, D.W. and Drikell, J.A. 2001. Carotenoid content of selected Indonesian fruits. J. Food Comp. Anal. 14, 169-176

Setyadjit and Murtiningsih 1990. The effect of arrangement and use of matresses on the damage of salak during transport (in Indonesian). Penel. Hort. 4, 1-7

Sctyadjit and Sjaifullah 1993. Study on several important parameters for designing modified atmosphere storage of *Salacca* Bali fruit (in Indonesian). Penel. Hort. 5 (3), 79-85

Seymour, G.B., Taylor, J.E. and Tucker, G.A. 1993. Biochemistry of fruit ripening. Chapman & Hall. London, UK

Shaner, D.L. and Boyer, J.S. 1976. Nitrate reductase activity in maize (*Zea mays* L.) leaves: 1. Regulation by nitrate flux. Plant. Physiol. 58, 623-636

Shewfelt, R. 1999. What is quality? Postharvest Biol. Technol. 15, 197-200

Siswandono 1995. The development of quality and quantity of salak pondoh fruit (*Salacca edulis* Reindw.) (in Indonesian). Prosiding Simposium Hortikultura Nasional. Perhimpunan Hortikultura Indonesia dan Universitas Brawijaya. Malang, Indonesia, 204-209

Soedibyo and Poernomo 1973. Packaging and storage of salak from Bali (in Indonesian). Hasil Penelitian Hortikultura. Lembaga Penelitian dan Pengembangan Hortikultura. Departemen Pertanian. Jakarta, Indonesia

Soleh, M., Ernawanto, Q.D., Wijadi, R.D. and Soemarsono, S.R. 1993. The effect of leaf pruning and fertilising on growth and production of salak (in Indonesian). Laporan Penelitian Sub Balai Hortikultura Malang, Departemen Pertanian, Jawa Timur, Indonesia, 46-63

Soleh, M., Suhardjo and Suryadi, A. 1995. The effect of watering and macro and micro fertilising to salak production (in Indonesian). Laporan Hasil Penelitian. Sub Balihorti Malang. Departemen Pertanian. Malang, Indonesia

Sosrodihardjo, S. 1986. Physical and chemical development of salak pondoh (in Indonesian). Bul. Penel. Hort., 13 (2), 54-62

Steel, R.G.D., Torrie, J..H. and Dickey, D. 1997. Principles and procedures of statistics. McGraw-Hill, New York, USA

Stoldt, W. 1952. Vorschlag zur Vereinheitlichung der Fettbestimmung in Lebensmitteln. Fette u. Seifen, 54, 206-207

Stoneman, G.L. and Dell, B. 1993. Growth of *Eucalyptus marginata* (jarrah) seedlings in a greenhouse in response to shade and soil temperature. Tree Physiol. 13, 239-252

Stoneman, G.L., Dell, B. and Turner, N.C. 1995. Growth of *Eucalyptus marginata* (jarrah) seedlings in Mediterranean-climate forest in south-west Australia in response to overstorey, site and fertiliser application. For. Ecol. Manage. (79), 173-184

Subagyo, Marsoedi dan Karama, S., 1996. The prospect of the development of peat land for agriculture (in Indonesian). Paper presented at "Seminar pengembangan teknologi berwawasan lingkungan untuk pertanian pada lahan gambut", 26 September, Bogor, Indonesia

Sudaryono, T., Prahardini, P.E.R., Purnomo, S. and Soleh, M. 1992. Distribution and estimation of germplasm collection and salak arrangement based on the izosyme analysis (in Indonesian). Sub Balai Penelitian Hortikultura Malang. Departemen Pertanian. Jawa Timur, Indonesia

Sudaryono, T., Purnomo, S. and Soleh, M. 1993. Cultivar distribution and estimation of area development of *Salacca* (in Indonesian). Penel. Hort. 5, 1-4

Sudaryono, T. and Soleh, M. 1994. Root induction in vegetative propagation of salak (in Indonesian). Penel. Hort. 6 (2), 13–17

Sudaryono, T. 1995. Results of salak research (in Indonesian). Prosiding Evaluasi Hasil Penelitian Hortikultura dalam Pelita V. Pusat Penelitian dan Pengembangan Hortikultura. Departemen Pertanian. Jakarta, Indonesia

Sudaryono, T., Martono and Saádah, S.Z. 1997. Technological assessment of *Salacca* mother-trees management (in Indonesian). Prosiding Seminar Hasil Penelitian/Pengkajian. Sub Balai Penelitian Hortikultura Malang. Departemen Pertanian. Jawa Timur, Indonesia

Sudaryono, T., Rosmahani, L., Suryadi, A., Ernawanto, Q.D. and Srihastuti, E. 1999. Technology packaging for increasing the harvest frequency of high quality *Salacca* from East Java (in Indonesian). Prosiding Seminar Hasil Penelitian/Pengkajian BPTP Karangploso. Departemen Pertanian. Indonesia, 122-128

Suhardi 1995. Changes in sugar and starch contents of salak pondoh during fruit maturation (in Indonesian). Agritech 10 (1,2,3), 10-13

Suhardi 1997. Respiration of salak pondoh after harvest (in Indonesian). Bul. II. Instiper 5 (2), 18-25

Suhardi, Tranggono and Santosa, U. 1997. Chemical and sensory changes of salak pondoh during modified atmosphere storage (in Indonesian). Agritech 17 (1), 6-9

Suhardjo, Wijadi, R.D. and Manan, K.A. 1995. The effect of harvesting stage on quality change of salacca cv pondoh during room storage (in Indonesian). Penel. Hort. 7(1), 62-71

Sulandra, K., Jamasuta, G.P., Agung, G.N., Buda, K. and Utama, I.B.D. 1987. The effect of packaging and fungicides on salak fruit quality during transport and storage (in Indonesian). Laporan Penelitian. Universitas Udayana. Bali, Indonesia

Sulladmath, V.V. 1983. Accumulation of mineral elements at different stages in developing sapodilla fruit. Sci. Hort. 19 (1-2), 79– 83

Sumardi, I., Sutikno and Susanti, S. 1995. *In vitro* preservation in salak pollen (*Salacca edulis* Reinw.) (in Indonesian). Biologi, 1(10), 445–451

Sunarmani. 1988. Study on the effect of soaking of salak in natriumbisulfid on the characteristics of salak syrup (in Indonesian). Master Thesis. Program Pasca Sarjana. Universitas Gadjah Mada. Yogyakarta, Indonesia

Supriyadi, Suhardi, Suzuki, M., Yoshida, K., Muto, T., Fujita, A. and Watanabe, N. 2002. Changes in the volatile compounds and in the chemical and physical properties of snake fruit (*Salacca edulis* Reinw.) cv Pondoh during maturation. J. Agric. Food Chem. 50, 7627–7633

Susanti, S. 1996. Ultrastructure of salak pollen (*Salacca edulis* Reinw.) before and after storage (in Indonesian). Laporan Penelitian Fakultas Biologi, Universitas Gadjah Mada, Yogyakarta, Indonesia

Susanti, S. 1997. Chemical change of the pollen wall of salak (*Salacca edulis* Reinw.) after storage (in Indonesian). Biologi 2 (7), 343–353

Suter, I.K. 1988a. Study on the characteristics of salak fruits in Bali as basis for the establishment of the product quality (in Indonesian). Disertasi Program Doktor. Fakultas Pasca Sarjana. Institut Pertanian Bogor. Bogor, Indonesia

Suter, I.K. 1988b. Postharvest physiological characteristics of some salak cultivars. Laporan Penelitian (in Indonesian). Fakultas Pertanian. Universitas Udayana. Bali, Indonesia

Termaat, A., Passioura J.B. and Munns, R. 1985. Shoot turgor does not limit shoot growth of NaCl-affected wheat and barley. Plant Physiol. 77, 869-872

Thamrin, M. 1998. The effect of organic soil conditioner and minerals on the productivity of salak pondoh in Sleman district (in Indonesian), Prosiding Seminar dan Lokakarya, Departemen Pertanian, Indonesia, 191-197

Thamrin, M., Djaafar, T.F. and Mudjisihono, R. 1988. Research and study on the increase of fruit quality of salak pondoh for export purpose through pre- and postharvest technology in Sleman district of Yogyakarta province (in Indonesian). Laporan Penelitian, Badan Penelitian dan Pengembangan Pertanian, Departmen Pertanian. Yogyakarta, Indonesia

Thomas, D.S. and Turner, D.W. 2001. Banana (*Musa* sp.) leaf gas exchange and chlorophyll fluorescence in response to soil drought, shading and lamina folding. Sci. Hort. 90, 93-108

Tieman, D.M. and Handa, A.K. 1994. Reduction in pectin methylesterase activity modifies tissue integrity and cation level in ripening tomato (*Lycopersicon esculentum* Mill.) fruits, Plant Physiol. 106, 429 – 436

Tjahjadi, N. 1988. Cultivation of salak (in Indonesian), 5th edn. Penerbit Kanisius, Yogyakarta, Indonesia

Tranggono 1998. Respiration pattern and flavour compound during maturation of salak pondoh (in Indonesian). Agritech. 18 (2), 1-4

Tu, N. T. M., Onishi, Y., Choi, H. S., Kondo, K., Bassore, S. M., Ukeda, H. and Sawamura, M. 2002. Characteristic odor components of *Citrus sphaerocarpa* Tanaka (Kabosu) cold-pressed peel oil. J. Agric. Food Chem. 50, 2908-2913

Tucker, G.A. 1993. Introduction. *In*: Seymour, G.B., Taylor, J.E. and Tucker, G.A. (eds.). Biochemistry of fruit ripening, Chapman & Hall, London, UK, 1-51

Turner, N.C. 1979. Drought resistance and adaptation to water deficits in crop plants. *In*: Mussel, H. and Staples, R. (Eds). Stress physiology in crop plants. Wiley-Interscience, New York, USA, 181-194

Ulrich, R. 1970. Organic acids. *In*: Hulme, A.C. (ed.). The biochemistry of the fruit and their products, Vol. 1, Academic press, London, UK, 89-118

Ulrichs, C. 1999. Dynamik des Kohlenhydratmetabolismus in der Nachernte am Beispiel von *Daucus carota* L. ssp. *sativus* (Hoffm.), Dissertation Humboldt-Universität Berlin, Der Andere Verlag, Osnabrück, Germany

User contributed software 'sniff.exe': http://www.chem.agilent.com/cag/servsup/usersoft/main.html.

Valim M. F., Rouseff, R. L. and Lin, J. 2003. Gas chromatographic-olfactometry characterization of aroma compounds in two types of cashew apple nectar. J. Agric. Food Chem. 51, 1010-1015

Van Heel, W.A. 1977. On the morphology of the ovules in *Salacca* (Palmae). Blumea 23, 371-375

von Caemmerer, S. and Farquhar, G.D. 1981. Some relationship between the biochemistry of photosynthesis and the gas exchange of leaves. Planta 153, 376-387

Vu, J.C.V. 1999. Photosynthetic responses of citrus to environmental changes. *In*: Pessarakli, M. (ed.). Handbook of plant and crop stress (2nd ed.). Marcel Dekker, Inc. New York, and Basel, 947 – 961

Walsh, L.M. and Murdoch, J.T. 1963. Recovery of fixed ammonium by corn in greenhouse studies. Soil Sci. Soc. Am. Proc. 27: 200-204

Watada A. E., Herner, R. C., Kader, A. A., Remani, R. J. and Staby, G. L. 1984. Terminology for the description of developmental stages of horticultural crops. Hort. Sci. 19 (1), 20-21

Watzl, B. and Leitzmann, C. 1999. Bioaktive Substanzen in Lebensmitteln. 2. Auflage. Hippokrates, Stuttgart, Germany

Westphal, E. and Jansen, P.C.M. 1989. Plant resources of South-East Asia. A Selection. Pudoc. Wageningen. The Netherlands

Whiting, G.C. 1970. Sugars. *In*: Hulme, A.C. (ed.). The biochemistry of the fruit and their products, Vol. 1, Academic Press. London, UK, 1-31

Whitmore, T.C. 1973. Palms of Malaya. Oxford University Press, London. UK

Wijadi, D. and Suhardjo 1992. Effect of packaging on damage of salak fruit during transport (in Indonesian). Laporan Hasil Penelitian Hortikultura Malang. Departemen Pertanian. Jawa Timur, Indonesia

Wijadi, D. and Suhardjo 1994. Processing technique for salak fruit (in Indonesian). Prosiding Seminar Hasil Penelitian Buah-buahan 1993/1994. Sub Balai Penelitian Hortikultura Malang. Departemen Pertanian. Jawa Timur, Indonesia

Wijana, G., Gunadi, I.G.A. and Putra, N.K. 1993. Increasing yield quantity and quality of Balinese snake fruit by determination of fruit thinning time and the number of fruits per bunch (in Indonesian). Laporan Penelitian. Fakultas Pertanian Universitas Udayana. Denpasar. Bali, Indonesia

Wijaya, C.H., Ulrich, D., Lestari, R., Schippel, K. and Ebert, G. 2005. Identification of potent odorants in different cultivars of snake fruit (*Salacca zalacca* (Gaertn.) Voss) using gas chromatography-olfactometry. J. Agric. Food Chem. In press

Wills, R.H.H., Lee, T.H., Graham, D., McGlasson, W.B. and Hall, E.G. 1981. Postharvest, an introduction to the physiology and handling of fruit and vegetables, New South Wales Univ. Press, Australia

Winarno, M. 1997. The policy on developing horticulture (in Indonesian). Panitia Dies Natalis ke–51. Fakultas Pertanian. Universitas Gadjah Mada. Yogyakarta, Indonesia

Winarno, M. 2000. Horticulture prospect and plant diversity (in Indonesian). Paper presented at "Hari cinta puspa dan satwa nasional". Bogor, Indonesia

Wong, K.C. and Tie, D.Y. 1993. Volatile constituents of salak (*Salacca edulis* Reinw.) fruit. Flav. Fragr. J. 8, 321-324

Wong, S.C., Cowan, I.R. and Farquhar, G.D. 1978. Leaf conductance in relation to assimilation in *Eucalyptus pauciflora* Sieb. Ex Spreng. Influence of irradiance and partial pressure of carbon dioxide. Plant Physiol. 62, 670-674

Wrasiati, L.P. 1997. Bee wax coating to maintain quality of Balinese *Salacca* (in Indonesian). Master Thesis. Fakultas Pasca Sarjana. Universitas Gadjah Mada. Yogyakarta, Indonesia

Wuryani, S. 1999. A simulation model of shelf life prediction for edible coated minimally processed fruits under atmospheric conditions: Arumanis mango and pondoh salak (in Indonesian). Master Thesis. Program Pasca Sarjana. Institut Pertanian Bogor. Bogor, Indonesia

ZMP-Markbilanz Obst. 2000. Deutschland. Europaeische Union. Weltmarkt. ZMP Zentrale Mark-und Preisberichtstelle GmbH. Bonn, Germany